Lecture Notes in Chemistry

Edited by:

Prof. Dr. Gaston Berthier
Université de Paris

Prof. Dr. Michael J. S. Dewar
The University of Texas

Prof. Dr. Hanns Fischer
Universität Zürich

Prof. Dr. Kenichi Fukui
Kyoto University

Prof. Dr. George G. Hall
University of Nottingham

Prof. Dr. Jürgen Hinze
Universität Bielefeld

Prof. Dr. Joshua Jortner
Tel-Aviv University

Prof. Dr. Werner Kutzelnigg
Universität Bochum

Prof. Dr. Klaus Ruedenberg
Iowa State University

Prof Dr. Jacopo Tomasi
Università di Pisa

Sándor Fliszár

Atoms, Chemical Bonds and Bond Dissociation Energies

ringer-Verlag Berlin Heidelberg GmbH

Author

Sándor Fliszár
Département de Chimie
Faculté des Arts et Sciences
Université de Montréal
Case postale 6128
Succursale centre-ville
Montréal (Québec), Canada H3C 3J7

ISBN 978-3-540-58237-3   ISBN 978-3-642-51492-0 (eBook)
DOI 10.1007/978-3-642-51492-0

Cip data applied for

This work is subject to copyright. All rights are reserved, whether the whole or part
of the material is concerned, specifically the rights of translation, reprinting, re-use
of illustrations, recitation, broadcasting, reproduction on microfilms or in any other
way, and storage in data banks. Duplication of this publication or parts thereof is
permitted only under the provisions of the German Copyright Law of September 9,
1965, in its current version, and permission for use must always be obtained from
Springer-Verlag Berlin Heidelberg GmbH. Violations are liable for prosecution
under the German Copyright Law.

© Springer-Verlag Berlin Heidelberg 1994
Originally published by Springer-Verlag Berlin Heidelberg New York in 1994

Typesetting: Camera ready by author
SPIN: 10473116   51/3140 - 543210 - Printed on acid-free paper

Preface

Chemical bonds are at the heart of this *Lecture Notes in Chemistry*. The whole story, i.e., the theory and its applications, is the product of twenty or so years' work. This seems a lot of time, but we did not know, then, into what sort of adventures we would be drawn by such a simple question as: 'what is the energy of a chemical bond?' Now we know.

It all began like a jig-saw puzzle. Now, as we all would agree, the truly elegant way of assembling a puzzle wants us to start at the upper left-hand corner and to proceed methodically from left to right, line by line, until completion of the picture, but this is hardly how things are done in real life. We did it the 'usual way', by piecing together, here and there, the bits as we recognized them, thus obtaining patches revealing fragments of the total picture. Intuition was as good as deductive reasoning when selecting the topics supposedly leading to a consistent theory of chemical bonds, but only patient processing of general theory permitted the bridging of gaps and filling in of the blanks to produce a theory that unfolds logically from alpha to omega, as if it had always existed in this form. Thus we hope, in presenting this work, to offer a 'complete' story for the reader's benefit, but do it in a way that should keep him on his toes when it comes to questions that deserve additional investigations: surely, there is room for improvements.

The formula for the energy of a chemical bond in a ground-state molecule —the so-called *'intrinsic' bond energy*—translates intuitive expectations, namely, that the energy of a bond formed by atoms k and l should depend on the amount of electronic charge carried by these atoms. The underlying physical picture is that of 'valence electrons' in the field of 'effective' nuclear charges, a picture best explained for isolated atoms (Chapter 1). Chapter 2 adapts the results thus obtained to molecular systems and Chapter 3 goes on with a straightforward demonstration of the bond-energy formula. Next we consider *bond dissociation energies* (Chapter 4) and derive an expression relating dissociation and intrinsic bond energies. Just like the latter, bond dissociation energies also depend on the charges carried by the bond-forming atoms. This is a good thing for us to know because, if we figure out how the environment of a molecule affects its charge distribution—which is a matter of carrying out population analyses—we can learn how the dissociation of chemical bonds found in a molecule can be affected by its environment. Of course, whether we consider intrinsic bond energies or bond dissociation, the key is in the evaluation of physically meaningful atomic charges—definitely a considerable task in itself (Chapter 5). Numerical applications and comparisons with experiment are indicated in Chapter 6, followed by a brief assessment (Chapter 7). Finally, useful information facilitating comparisons between theory and experiment is given in the Appendix.

Let us not forget that what is offered here is just *our* contribution to the understanding of chemical bonds, i.e., much less than what is generally known about them. References for further reading are given where appropriate: their conciseness should not mislead anyone as for their importance. All the numerical examples were newly recalculated for this *Lecture Notes* and may here and there differ from those of the original literature. These minor differences do not obscure what we hope is an inkling of truth worth thinking about: the charge dependence of bond energies.

Dedication and Acknowledgements

I wish to dedicate this work to my students and post-doctoral fellows who made things possible: Marie–Thérèse Béraldin, Jacques Bridet, Jean–Louis Cantara, Guy Cardinal, Michel Comeau, Geneviève Dancausse, Normand Desmarais, Aniko Foti, Marielle Foucrault, Annick Goursot, Jacques Grignon, Hervé Henry, Gérard Kean, Claude Mijoule, Camilla Minichino, Andrea Peluso, François Poliquin, Réal Roberge, and Edouard Vauthier. I also wish to express my heartfelt thanks for active help, advice, patience and friendship to Professors Giuseppe Del Re, Vincenzo Barone, Jean–Marie Leclercq, Simone Odiot, and Dennis Salahub. Last but not least, I include Signora Dora and Don Gaetano Lampo in this dedication. These fine people made me a better man by teaching me important things, such as the *true* tolerance that is so uniquely part of Neapolitan culture. They also made me a little, but just a little, fatter.

Sándor Fliszár

Contents

Chapter 1

Core and Valence Regions of Atoms

The idea of subdividing an atom into an inner core and an outer valence region is not a new one. Though intuitively appealing— much of the chemistry resides in the valence region—such a core–valence separation is by no means obvious. Indeed, no electron can be assigned to any region in particular because each individual electron occupies the entire atom. Fortunately, we can benefit from the notion of *stationary* electron density. The particles are not at rest, but the probability density does not change with time. We look for a stationary, physically meaningful core–valence partitioning of atoms, for future use in applications to molecules.

1.1 The Hartree–Fock approximation

The Hartree–Fock self-consistent-field (SCF) method is the primary tool used in this Chapter. It is rooted in the time-independent one-electron Schrödinger equation

$$\left[-\frac{1}{2}\nabla^2(1) + V(r_1) \right] \phi_i(1) = \epsilon_i(1)\phi_i(1) . \tag{1.1}$$

The effective one-electron operator indicated in brackets includes the kinetic-energy operator, $-\frac{1}{2}\nabla^2$, and an effective potential energy, $V(r_1)$, taken as an averaged function of r_1— the distance of electron 1 from the nucleus. In this approximation, electron 1 moves in the field created by both the nuclear charge Z and a smeared-out static distribution of electric charge due to electrons $2, 3, \ldots, n$. The eigenfunction $\phi_i(1)$ is a one-electron orbital and $\epsilon_i(1)$ is the corresponding energy.

The Hartree–Fock equation

$$\hat{F}\phi_i = \epsilon_i\phi_i \tag{1.2}$$

has the same form as (1.1) but introduces spin explicitly in the description of the wave function. The ϕ_i's are now spin-orbitals and ϵ_i is the eigenvalue of spin-orbital i. The effective Hartree–Fock Hamiltonian, $\hat{F}$, contains two one-electron operators, namely the kinetic-energy operator

$$\hat{T} = -\frac{1}{2}\nabla^2 \tag{1.3}$$

and the potential energy for the attraction between the electron and the nucleus of charge Z, i. e.

$$\hat{V}_{\text{ne}} = -\frac{Z}{r} . \tag{1.4}$$

In addition, $\hat{F}$ contains two bielectronic operators. They describe the interaction between the electron occupying spin-orbital i and the other electrons found in the atom. Namely, for the interaction between electrons 1 and 2 at a distance r_{12}, we have the Coulomb operator $\hat{J}_j$ and the exchange operator $\hat{K}_j$ defined by

$$\hat{J}_j(1)\phi_i(1) = \phi_i(1) \int |\phi_j(2)|^2 \frac{1}{r_{12}} \, d\tau_2 \tag{1.5}$$

$$\hat{K}_j(1)\phi_i(1) = \phi_j(1) \int \frac{\phi_j^*(2)\phi_i(2)}{r_{12}} \, d\tau_2 \tag{1.6}$$

where $d\tau$ is the volume element

$$d\tau = r^2 \sin\theta \, d\theta \, d\varphi \, dr \tag{1.7}$$

and the subscript "2" refers to electron 2. Of course, similar Coulomb and exchange operators for electrons $3, \ldots, n$ are also part of $\hat{F}$. The Coulomb integral $\int \phi_i^*(1)\hat{J}_j(1)\phi_i(1) \, d\tau_1$ represents the repulsion between electron 1 and a smeared-out electron with density $|\phi_j(2)|^2$. The exchange integral $\int \phi_i^*(1)\hat{K}_j(1)\phi_i(1) \, d\tau_1$ arises from the requirement that the wave function be antisymmetric with respect to electron exchange in order to satisfy Pauli's indistinguishability principle of identical particles. Because of the occurrence of ϕ_j terms in the Coulomb and exchange operators and, thus, in $\hat{F}$, eq. (1.2) must be solved iteratively until self-consistency is achieved, the result being a set of self-consistent (SCF) eigenfunctions ϕ_{1s}, ϕ_{2s},..., with orbital eigenvalues ϵ_{1s}, ϵ_{2s},...

These Hartree–Fock energies are reasonably good approximations to the orbital energies of an atom, as determined by X-ray and optical spectroscopy term values. We take them as the energies of the individual electrons knowing, of course, that each of these orbital energies is computed over the entire atom and that no localization of the individual electrons in certain regions of space should be attempted. The energy ϵ_i of an electron in orbital ϕ_i consists of its kinetic energy

$$T_i = \int \phi_i^* \hat{T} \phi_i \, d\tau \tag{1.8}$$

and of its potential energy. The interaction of that electron with the nucleus is, from eq. (1.4),

$$V_{ne,i} = -Z \int \frac{|\phi_i(r)|^2}{r} \, d\tau. \tag{1.9}$$

On the other hand, the interaction of that electron, called here electron 1, with electrons 2, 3,..., n is given by the appropriate sums of Coulomb and exchange integrals, e.g., for electron 2 interacting with electron 1,

$$J_{12} = \iint |\phi_i(1)|^2 \frac{1}{r_{12}} |\phi_j(2)|^2 \, d\tau_1 \, d\tau_2 \tag{1.10}$$

$$K_{12} = \iint \frac{\phi_i^*(1)\phi_j(1)\phi_j^*(2)\phi_i(2)}{r_{12}} \, d\tau_1 \, d\tau_2 \tag{1.11}$$

which are conveniently carried out with the help of the following expansion [1] of $1/r_{12}$ in terms of spherical harmonics

$$\frac{1}{r_{12}} = \sum_{l=0}^{\infty} \sum_{m=-l}^{l} \frac{4\pi}{2l+1} \frac{r_<^l}{r_>^{l+1}} [Y_l^m(\theta_i, \varphi_i)]^* Y_l^m(\theta_j, \varphi_j) \tag{1.12}$$

where $r_>$ is the larger and $r_<$ is the smaller of r_1 and r_2. The total repulsive potential energy experienced by electron 1 in orbital ϕ_i is thus computed from a sum of Coulomb integrals, $J_{12} + J_{13} + \cdots$, from which we subtract the exchange integrals involving the electrons with the same spin as electron 1. Briefly, ϵ_i consists of the kinetic energy of an electron in orbital ϕ_i, plus the potential energy of interaction between that electron and the nucleus and all the other electrons.

Consider now the normalized Hartree–Fock spatial orbitals ϕ_{1s}, ϕ_{2s},..., with energies ϵ_{1s}, ϵ_{2s},..., respectively, occupied by ν_i ($= 0, 1$, or 2) electrons.

The potential energy of an electron with energy ϵ_i includes, in an average way, the repulsion between this electron and all the other electrons. The sum of the orbital energies, $\sum_i \nu_i \epsilon_i$, thus counts each interelectronic repulsion twice. The Hartree–Fock energy of the atom, E, is therefore

$$E = \sum_i \nu_i \epsilon_i - V_{ee} \tag{1.13}$$

where V_{ee} is the interelectronic repulsion computed over the entire atom.

It is understood that all the integrals considered in this Section are definite integrals over the full range of all the coordinates. Let us now examine what happens during the partitioning of an atom into inner and outer atomic regions.

1.2 Inner and outer atomic regions

Consider the partitioning of an atom (or ion) into two regions of space: a spherical inner region of radius r_b, centered at the nucleus, and an outer region, extending from r_b to infinity. The Hartree–Fock equation (1.2) is our starting point. Multiplication from the left by ϕ_i^*, integration from r_b to ∞, and summation over all occupied orbitals i leads to

$$\sum_i \int_{r_b}^{\infty} \nu_i \phi_i^* \hat{F} \phi_i \, d\tau = \sum_i \int_{r_b}^{\infty} \nu_i \phi_i^* \epsilon_i \phi_i \, d\tau \tag{1.14}$$

where ν_i is the occupation of the normalized orbital with eigenvalue ϵ_i. No special meaning is attached to any particular radius r_b defining the boundary surface separating the inner and outer regions: r_b can be chosen at will, whether or not this boundary corresponds to a physically meaningful core–valence separation. We shall learn that the validity of (1.14) depends on whether or not the ϕ_i's are true Hartree–Fock orbitals: approximate solutions do not satisfy (1.14) for all values of r_b.

Now we transform (1.14) into something more practical. Let us begin with its right-hand side. The integral

$$N_i^v = \nu_i \int_{r_b}^{\infty} \phi_i^* \phi_i \, d\tau \tag{1.15}$$

represents the number of electrons of orbital i found in the outer region (which we call for simplicity, but admittedly somewhat sketchily, valence region; hence the superscript "v"). Equation (1.14) thus becomes

$$\sum_i \int_{r_b}^{\infty} \nu_i \phi_i^* \hat{F} \phi_i \, \mathrm{d}\tau = \sum_i N_i^v \epsilon_i \,. \tag{1.16}$$

Next, we proceed with the left-hand side of (1.16). The evaluation of the monoelectronic integrals occurring in it is straightforward. The nuclear-electronic potential energy of the $\sum_i N_i^v$ outer electrons in the field of the nuclear charge Z is, from (1.9),

$$V_{\mathrm{ne}}^v = -Z \sum_i \nu_i \int_{r_b}^{\infty} 4\pi r \, |\phi_i(r)|^2 \, \mathrm{d}r \tag{1.17}$$

and their kinetic energy is, from (1.8),

$$T^v = \sum_i \nu_i \int_{r_b}^{\infty} \phi_i^* \hat{T} \phi_i \, \mathrm{d}\tau \,. \tag{1.18}$$

Now we turn to the bielectronic integrals appearing in the left-hand side of (1.16). They require a little attention. Consider the Coulomb operator $\hat{J}_j$ (1.5). The integral $\int \ldots \mathrm{d}\tau_2$ is over all space, but can be separated into two contributions, namely, in short-hand notation,

$$\int^{\tau_2^c} \ldots \mathrm{d}\tau_2 + \int^{\tau_2^v} \ldots \mathrm{d}\tau_2$$

where the first integral (from 0 to r_b) covers the inner (core) region τ_2^c, whereas the second one (from r_b to ∞) is for the outer (valence) space τ_2^v. Using (1.5), the integration between r_b and ∞ required by (1.16) thus gives the following Coulomb repulsion:

$$\int^{\tau_1^v} \phi_i^*(1) \hat{J}_j(1) \phi_i(1) \, \mathrm{d}\tau_1 = \int^{\tau_1^v} \int^{\tau_2^c} \ldots \mathrm{d}\tau_2 \, \mathrm{d}\tau_1 + \int^{\tau_1^v} \int^{\tau_2^v} \ldots \mathrm{d}\tau_2 \, \mathrm{d}\tau_1. \tag{1.19}$$

These integrals describe a potential energy of interaction involving the part of electron 1 assigned to the valence space τ^v and a smeared-out electron with density $|\phi_j(2)|^2$. The two terms of the right-hand side of (1.19) differ from one another because the first integral 'sees' only the part of $|\phi_j(2)|^2$ found in the core space τ^c, whereas the second integral concerns only the part of $|\phi_i(2)|^2$ confined within the valence region τ^v.

At last, we can carry out the sum indicated in (1.16). This sum also includes a term arising from $\hat{J}_i(2)\phi_j(2)$, whose integral is evaluated as just shown for $\hat{J}_j(1)\phi_i(1)$, i.e.,

$$\int^{\tau_2^{\rm v}} \phi_j^*(2)\hat{J}_i(2)\phi_j(2)\,{\rm d}\tau_2 = \int^{\tau_2^{\rm v}}\int^{\tau_1^{\rm c}} \ldots {\rm d}\tau_1\,{\rm d}\tau_2 + \int^{\tau_2^{\rm v}}\int^{\tau_1^{\rm v}} \ldots {\rm d}\tau_1\,{\rm d}\tau_2.$$

As a consequence, the summation over all occupied orbitals counts once again the 1–2 Coulomb interaction confined within the valence space, i.e., the $\int^{\tau_1^{\rm v}}\int^{\tau_2^{\rm v}} \ldots {\rm d}\tau_2\,{\rm d}\tau_1$ term already found in (1.19), and adds the new integral

$$\int^{\tau_2^{\rm v}}\int^{\tau_1^{\rm c}} \ldots {\rm d}\tau_1\,{\rm d}\tau_2$$

describing the interaction between the part of electron 2 found in $\tau^{\rm v}$ and the part of electron 1 found in the core region. Briefly, all interactions involving exclusively electron densities assigned to the outer space $\tau^{\rm v}$ are counted twice in the sum (1.16), whereas the cross-interactions between electrons assigned to $\tau^{\rm v}$ and those assigned to $\tau^{\rm c}$, represented by the appropriate $\int^{\tau_1^{\rm v}}\int^{\tau_2^{\rm c}} \ldots {\rm d}\tau_2\,{\rm d}\tau_1 + \int^{\tau_2^{\rm v}}\int^{\tau_1^{\rm c}} \ldots {\rm d}\tau_1\,{\rm d}\tau_2$ integrals, are counted only once. The same reasoning applies to the exchange operator (1.6) and to its integral $\int \ldots {\rm d}\tau_1$, followed by summation over i. Down the line, taking now all the appropriate Coulomb and exchange integrals into account, it is seen that the quantity

$$V_{\rm ee}^{\rm vv} = \text{electron} - \text{electron repulsion involving only}$$
$$\text{the electrons of the outer (valence) region}$$

is counted twice in the left-hand side of (1.16), whereas

$$V_{\rm ee}^{\rm cv} = \text{repulsion between the valence electrons}$$
$$\text{and those of the inner (core) region}$$

is counted only once.

The final result, including now $V_{\rm ne}^{\rm v}$ and $T^{\rm v}$, eqs. (1.17) and (1.18), respectively, is therefore

$$\sum_i N_i^{\rm v}\epsilon_i = T^{\rm v} + V_{\rm ne}^{\rm v} + 2V_{\rm ee}^{\rm vv} + V_{\rm ee}^{\rm cv}. \tag{1.20}$$

This equation [2, 3] is a handy form of (1.14). All the terms can be evaluated by standard procedures, using eqs. (1.8)–(1.12), simply by paying attention to the appropriate limits of integration. Evidently, all the quantities (except the ϵ_i's) are functions of the r_b of our choice, e.g., $N_i^{v} = N_i^{v}(r_b)$, $T^{v} = T^{v}(r_b)$, etc. A similar equation can also be derived for the core region, along the same lines, by carrying out the integration of $\phi_i^* \hat{F} \phi_i$ between 0 and r_b, i.e.,

$$\sum_i N_i^c \epsilon_i = T^c + V_{ne}^c + 2V_{ee}^{cc} + V_{ee}^{cv} \tag{1.21}$$

where T^c, V_{ne}^c, and V_{ee}^{cc} are the kinetic, the nuclear–electronic, and the interelectronic energies of the $\sum_i N_i^c$ core electrons. Note that (1.21) follows immediately from (1.20) and (1.13) because the individual core and valence terms add up to give the corresponding totals for the entire atom, i.e., $\sum_i N_i^c \epsilon_i + \sum_i N_i^v \epsilon_i = \sum_i \nu_i \epsilon_i$, $T = T^c + T^v$, $V_{ne} = V_{ne}^c + V_{ne}^v$, and $V_{ee} = V_{ee}^{cc} + V_{ee}^{vv} + V_{ee}^{cv}$.

Equations (1.20) and (1.21) are exact in Hartree–Fock theory. This means that true Hartree–Fock functions satisfy these equations for any r_b of our choice. Approximate wave functions, on the other hand, obtained from self-consistent-field calculations, only ensure this result for $r_b = 0$ because all integrals are carried out over the full range of coordinates, i.e., over the entire atom, in the SCF procedure. Let us now examine the orbital functions from that angle, using (1.20).

1.3   Hartree–Fock–Roothaan orbitals

Following Roothaan's proposal, the Hartree–Fock orbitals are usually represented as linear combinations of a set of known basis functions χ_k^{lm},

$$\phi_i = \sum_k c_{ki} \chi_k^{lm}. \tag{1.22}$$

This representation permits analytic calculations, as opposed to fully numerical solutions [4, 5] of the Hartree–Fock equation. Variational SCF methods using finite expansions (1.22) yield optimal analytic Hartree–Fock–Roothaan orbitals, and their corresponding eigenvalues, within the subspace spanned by the finite set of basis functions.

Commonly, one uses normalized Slater-type orbitals

$$\chi^{lm} = \frac{(2\zeta)^{n+1/2}}{[(2n)!]^n} \, r^{n-1} \, e^{-\zeta r} Y_l^m(\theta, \varphi) \tag{1.23}$$

where n is the principal quantum number and ζ is the orbital exponent. Alternatively, one can use (larger) linear combinations of Gaussian functions

$$N r^l e^{-\zeta r^2} Y_l^m(\theta, \varphi) \qquad (1.24)$$

which are particularly efficient in molecular calculations, in that they require less computer time than Slater integral evaluation. In atomic calculations, Slater functions are preferred because one-center integrals are no more difficult for Slater-type than for Gaussian-type orbitals and relatively few well-chosen Slater functions yield accurate results.

To get true Hartree–Fock orbitals, an infinite set of basis functions should be included in the expansion (1.22). The question is: how closely do finite expansions approach the true Hartree–Fock limit? Equation (1.20) offers an answer that does not require any *a priori* knowledge of results obtained at (or at least near) the Hartree–Fock limit.

Consider the difference, from (1.20),

$$D = \sum_i N_i^{\mathrm{v}} \epsilon_i - (T^{\mathrm{v}} + V_{\mathrm{ne}}^{\mathrm{v}} + 2V_{\mathrm{ee}}^{\mathrm{vv}} + V_{\mathrm{ee}}^{\mathrm{cv}}). \qquad (1.25)$$

This difference represents what we may call the 'noise' of a truncated expansion, with respect to a base line which is everywhere identically zero at the true Hartree–Fock limit. This difference is a function of r_b. For each r_b, there is one value of N^c, the number of electrons contained within the sphere of radius r_b. The use of D expressed as a function of N^c, i.e., $D = D(N^c)$, allows us to scan the entire atom, thus simplifying the presentation of results, by avoiding long tails for large r_b's. At the limits $N^c = 0$ and $N^c = N = $ total number of electrons, it is $D(N^c) = 0$ because $N^c = 0$ corresponds to the entire atom, for which the SCF result is correct, while for $N^c = N$ nothing is left in the outer region, meaning that all terms in (1.20) are naught.

Figures 1.1 and 1.2 are for argon and indicate how $D(N^c)$ varies with N^c. The results of Figure 1.1 describe the single-ζ and double-ζ Slater bases of Clementi and Roetti [6], those of Figure 1.2 describe Gaussian bases developed by Pople and coworkers [7]. Figure 1.3 displays Slater basis results for nickel. These examples are typical: the deviations calculated for the near-Hartree–Fock orbitals reported in [6] are barely visible but do exist[1].

Equation (1.25) measures how closely approximate orbital descriptions approach the true Hartree–Fock limit and thus provides a quality ordering of

---

[1] The noise of double-ζ calculations is roughly one tenth that of the single-ζ basis, while in near Hartree–Fock calculations this noise is further reduced by a factor of 10, approximately [2].

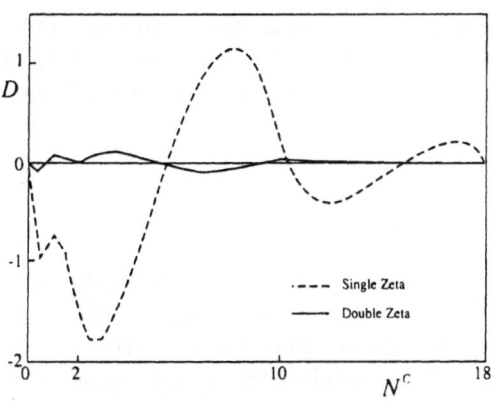

FIGURE 1.1. Argon. $D(N^c)$ calculated for single-ζ and double-ζ Slater bases, atomic units [2].

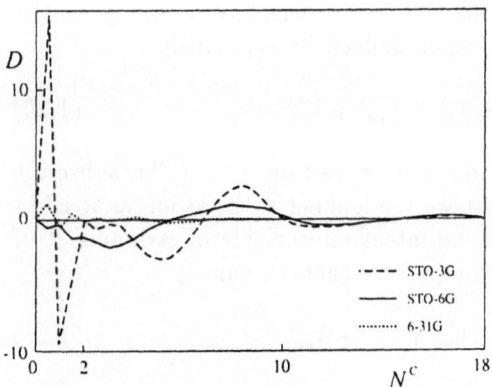

FIGURE 1.2. Argon. $D(N^c)$ calculated for Gaussian bases, atomic units [2].

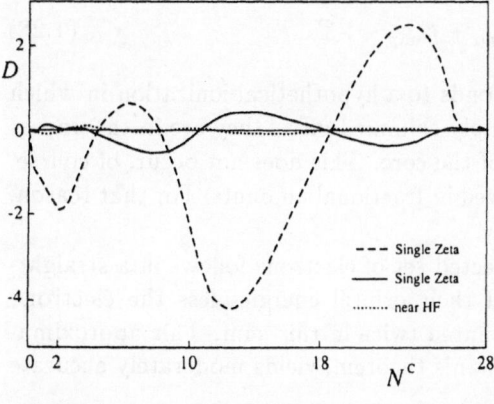

FIGURE 1.3. Nickel. $D(N^c)$ calculated for single-ζ, double-ζ and near-HF Slater bases, atomic units [2].

orbital bases with respect to one another and with respect to that limit, in a scale uniquely defined by the latter. It is now clear that numerical calculations involving (1.20) in one form or another require the use of near-Hartree–Fock quality wave functions.

1.4   The valence region energy

Two problems mingle at this point in time: what is a valence region? and what is its energy? Let us first eliminate what is *not* a valence region and what is *not* a valence region energy, keeping in mind that the latter should be an observable quantity and that a viable definition of valence region must be compatible with that observable.

Using eq. (1.20), we can define a 'valence region energy' $\sum_i N_i^v \epsilon_i - V_{ee}^{vv} = T^v + V_{ne}^v + V_{ee}^{vv} + V_{ee}^{cv}$ which is just the sum of the kinetic and potential energies associated with a valence region defined by r_b, namely

$$E_{unp}^v = T^v + V_{ne}^v + V_{ee}^{vv} + V_{ee}^{cv}. \tag{1.26}$$

E_{unp}^v restates the energy balance derived in Section 1.2. (The subscript "unp", for unperturbed, means that we are looking at an atom 'as it is' in the Hartree–Fock description, with no intent on our side of stripping it of its electrons.) Similarly, we can define a 'core energy', namely

$$E_{unp}^c = T^c + V_{ne}^c + V_{ee}^{cc} + V_{ee}^{cv}. \tag{1.27}$$

Remembering that $E = T + V_{ne} + V_{ee}$ is the energy of the entire atom, it follows from (1.26) and (1.27) that

$$E = E_{unp}^v + E_{unp}^c - V_{ee}^{cv}. \tag{1.28}$$

The valence energy (1.26) corresponds to a hypothetical ionization in which the valence electrons would be simply skimmed off as they are in the unperturbed atom, with no relaxation of the core. This does not occur, of course, because electrons cannot be removed in fractional amounts. For that reason, E_{unp}^v is not an observable quantity.

An observable energy for a selected set of electrons follows in a straightforward manner from the sum of their orbital energies less the electron-electron interactions which are counted twice in this sum. This approximation, reflecting the spirit of Koopman's theorem, yields moderately accurate results in Hartree–Fock theory[2]. The point is in the arbitrariness involved in the definition of core electrons: they can be selected to suit our convenience (e.g., we can freely select a two-electron $1s$ core for the first-row elements). It is clear that this approach simply amounts to an orbital-by-orbital separation, which is nothing new in Hartree–Fock theory. This method has merits of its own, of course: it lays the foundation of the so-called pseudopotential approximation [9] whereby inner-shell electrons are eliminated from the calculation and replaced by a suitably parametrized pseudopotential operator. But the concept of a physically meaningful core–valence separation melts away in this type of approach.

So we go back to the partitioning described in Section 1.2 and examine whether a valence energy other than (1.26) can be derived, complying with the requirement that this energy should represent a physical quantity. The problem at hand is best explained by an example. Suppose that there is some reason to assume that the core of carbon is adequately represented by a two-electron inner shell. In Hartree–Fock theory this core cannot consist of two pure $1s$ electrons: some $2s$ and $2p$ electron densities are found in this core, with some $1s$ density in the outer region. But upon ionization of the two $2s$ and the two $2p$ electrons, a C^{+4} ion is left behind with two pure $1s$ electrons. In other words, a relaxation is part of this process and is thus part of the description of the physical valence energy. Briefly, we have to look for a valence energy, E^v, by studying the general properties of an atom in which the core and valence regions are interdependent.

In this vein, we re-examine the problem with the understanding that in the event of an actual ionization one region benefits at the expense of the other from the occurrence of relaxation. The valence and core energies modified by this relaxation are not any longer those appearing in (1.28), but the modified valence-region energy, E^v, and its core counterpart, E^c, are still tied by this general energy balance, that is

$$E = E^v + E^c - V_{ee}^{cv}. \tag{1.29}$$

E^v corresponds to the energy required for the removal of the valence electrons, i.e., to minus the sum of the relevant ionization potentials. When n

---

[2] The two $4s$ and the six $4p$ electrons of krypton, for example, tentatively christened valence electrons, possess a total orbital energy of -5.4508 au. The Coulomb repulsion between these electrons is 13.4417 and the exchange is 0.8421 au, giving an interelectronic repulsion of 12.5996 au and a 'valence energy' of -18.0504 au. The sum of the corresponding ionisation potentials [8] is 18.6695 au. The agreement is as good as one can expect from an uncorrelated nonrelativistic calculation near the Hartree–Fock limit.

valence electrons are removed from a neutral ground-state atom A, an ion with charge $+n$ is left behind. The same ion is formed upon removal of $n+1$ electrons from a negative ion A^-, or upon removal of $n-1$ electrons from A^+. Briefly, the valence energies of A^-, A, A^+, etc. are always expressed by reference to the same final ion. Inspection of (1.29) indicates that the ground-state energy of this ion is

$$E^{ion} = E^c - V_{ee}^{cv}. \tag{1.30}$$

E^v and E^c are the unknowns of our problem. In the following we derive an expression for E^v without bothering about a physically valid core–valence separation, i.e., we treat E^v as if it were a continuous function of r_b. The acceptable discrete solutions of E^v are selected afterwards.

Consider first the electronic energy of a ground-state atom or ion with nuclear charge Z, $E = \langle \Psi | \hat{H} | \Psi \rangle$, and apply the Hellmann–Feynman theorem [10] taking Z as parameter. This gives, in conventional notation, at constant electron density ρ,

$$\left(\frac{\partial E}{\partial Z} \right)_\rho = \left\langle \Psi \left| \frac{\partial \hat{H}}{\partial Z} \right| \Psi \right\rangle = \left\langle \Psi \left| -\sum_i r_i^{-1} \right| \Psi \right\rangle$$

where the sum over i runs over all electrons. The well-known result is (1.9)

$$\left(\frac{\partial E}{\partial Z} \right)_\rho = \frac{V_{ne}}{Z}. \tag{1.31}$$

On the other hand, the virial theorem, $2E = V_{ne} + V_{ee}$, and the Hartree–Fock formula (1.13) combine to give

$$3E = V_{ne} + \sum_i \nu_i \epsilon_i. \tag{1.32}$$

This result and eq. (1.31) lead to

$$3E - Z \left(\frac{\partial E}{\partial Z} \right)_\rho = \sum_i \nu_i \epsilon_i. \tag{1.33}$$

We now define

$$\gamma = \frac{Z}{E} \left(\frac{\partial E}{\partial Z} \right)_\rho \tag{1.34}$$

and rewrite (1.33) as follows:

$$(3 - \gamma)E = \sum_i \nu_i \epsilon_i. \tag{1.35}$$

Finally, combining (1.31) and (1.34), we get

$$E = \frac{1}{\gamma} V_{\text{ne}}. \tag{1.36}$$

Equations (1.35) and (1.36) represent an identity in Hartree–Fock theory[3]. The particular interest offered by (1.36) lies in the fact that $\gamma = \frac{7}{3}$ appears to be the characteristic homogeneity of both Thomas–Fermi [11, 12, 13] and local density functional theory [14], in which case (1.35) gives the Ruedenberg approximation [15], $E = \frac{3}{2} \sum_i \nu_i \epsilon_i$, while (1.36) gives the Politzer formula [16], $E = \frac{3}{7} V_{\text{ne}}$. Equations (1.29), (1.35) and (1.36) are our working formulas. We pick up things where we left them with eq. (1.29) and write

$$\gamma E = \gamma^{\text{v}} E^{\text{v}} + \gamma^{\text{c}} (E^{\text{c}} - V_{\text{ee}}^{\text{cv}}). \tag{1.37}$$

This equation is more general than what would follow from a simple multiplication by γ. It is noncommittal as to whether the atomic γ suits the individual core and valence parts: in writing (1.37), γ is taken as the average of γ^{v} (with a weight of E^{v}) and γ^{c} (weighted by $E^{\text{c}} - V_{\text{ee}}^{\text{cv}}$), i.e.,

$$\gamma = \frac{\gamma^{\text{v}} E^{\text{v}} + \gamma^{\text{c}} (E^{\text{c}} - V_{\text{ee}}^{\text{cv}})}{E^{\text{v}} + E^{\text{c}} - V_{\text{ee}}^{\text{cv}}}.$$

Next we consider $V_{\text{ne}} = V_{\text{ne}}^{\text{v}} + V_{\text{ne}}^{\text{c}}$, which we write

$$V_{\text{ne}} = (V_{\text{ne}}^{\text{v}} + V_{\text{ee}}^{\text{cv}}) + (V_{\text{ne}}^{\text{c}} - V_{\text{ee}}^{\text{cv}}).$$

The reason for this association of terms is physical: in a central-force problem, V_{ne}^{v} and $V_{\text{ee}}^{\text{cv}}$ play roles that are similar in nature. V_{ne}^{v} measures the attraction of the valence electrons by the nucleus and $V_{\text{ee}}^{\text{cv}}$ the accompanying repulsion by the core electrons. This formulation is directly linked to the model adopted here, that of a charged inner sphere of radius r_{b} surrounded by valence electrons. With eq. (1.36) in mind we now write

$$\frac{1}{\gamma} V_{\text{ne}} = \frac{1}{\gamma^{\text{v}}} (V_{\text{ne}}^{\text{v}} + V_{\text{ee}}^{\text{cv}}) + \frac{1}{\gamma^{\text{c}}} (V_{\text{ne}}^{\text{c}} - V_{\text{ee}}^{\text{cv}}). \tag{1.38}$$

In this case $1/\gamma$ is the weighted average of $1/\gamma^{\text{v}}$ (with a weight of $V_{\text{ne}}^{\text{v}} + V_{\text{ee}}^{\text{cv}}$) and of $1/\gamma^{\text{c}}$ (with a weight of $V_{\text{ne}}^{\text{c}} - V_{\text{ee}}^{\text{cv}}$).

---

[3]The Hellmann–Feynman and virial theorems are satisfied by Hartree–Fock wave functions.

At this point we use eqs. (1.37) and (1.38) and solve for γ^v. After some algebra one obtains $(\gamma^v)^2 + b\gamma^v + c = 0$ with $b = -[(V_{ne}^v + V_{ee}^{cv})/E^v + \gamma]$ and $b^2 - 4c = [(V_{ne}^v + V_{ee}^{cv})/E^v - \gamma]^2$. The trivial root is $\gamma^v = \gamma$. The other root gives

$$E^v = \frac{1}{\gamma^v}(V_{ne}^v + V_{ee}^{cv}). \tag{1.39}$$

So far we have exploited eq. (1.36). We have one more step to go. Using (1.29) and (1.37) we write

$$(3 - \gamma)E = (3 - \gamma^v)E^v + (3 - \gamma^c)(E^c - V_{ee}^{cv})$$

and compare this expression with $\sum_i \nu_i \epsilon_i = \sum_i N_i^v \epsilon_i + \sum_i N_i^c \epsilon_i$. Equation (1.35) tells us that

$$\left[(3 - \gamma^v)E^v - \sum_i N_i^v \epsilon_i\right] + \left[(3 - \gamma^c)(E^c - V_{ee}^{cv}) - \sum_i N_i^c \epsilon_i\right] = 0.$$

This equation achieves a core–valence separation. The terms in brackets are certainly individually zero at the limits $N^c = 0$ and $N^v = 0$, but this does not warrant that these terms are individually zero for other values of N^c, i.e., that there is an N^c satisfying a meaningful core–valence separation. We shall tentatively proceed with

$$(3 - \gamma^v)E^v = \sum_i N_i^v \epsilon_i \tag{1.40}$$

and postpone momentarily the question about acceptable N^c values. Equation (1.40) defines E^v. It is the 'valence counterpart' of (1.35). Comparison with eq. (1.39) yields the energy formula [3]

$$E^v = \frac{1}{3}\left(V_{ne}^v + V_{ee}^{cv} + \sum_i N_i^v \epsilon_i\right) \tag{1.41}$$

which is visibly the valence counterpart of (1.32). E^v expresses a valence region energy that takes the relaxation of the core into proper account. It corresponds to the appropriate sum of ionization potentials. Using (1.20), eq. (1.41) becomes

$$E^v = \frac{1}{3}(T^v + 2V^v) \tag{1.42}$$

where $V^v = V_{ne}^v + V_{ee}^{vv} + V_{ee}^{cv}$ is the total potential energy of the $\sum_i N_i^v$ electrons associated with the outer (valence) region. For the entire atom, i.e., letting $N^c = 0$, eq. (1.42) reduces to $E = -T$ because $V/T = -2$ (virial theorem). While eq. (1.20) is valid for any r_b of our choice, E^v is meaningful only for discrete values of N^c.

In closing, let us compare this physically observable E^v with E^v_{unp}, eq. (1.26), describing the valence region of an unperturbed Hartree–Fock atom. We eliminate V^{vv}_{ee} from (1.26) with the help of (1.20) and compare the result with (1.41). This gives

$$E^v_{unp} = \frac{1}{2}(3E^v + T^v). \tag{1.43}$$

Surely, $E^v \neq E^v_{unp}$. Equation (1.43) indicates that the virial theorem, $-T^v = E^v$, and thus $-T^v = E^v_{unp}$, is not obeyed in the valence region.

Equations (1.41)–(1.43) still describe E^v as if it were a continuous function of r_b. There are restrictions, however, if E^v is ment to represent a physical quantity, namely a valence energy that measures the energy actually required for the removal of integer numbers of outer electrons.

1.5 Meaningful valence regions

The question is: is it possible to define a meaningful core–valence separation? Formal Hartree–Fock theory does not speak out on this topic. But numerical Hartree–Fock results offer vivid guidelines. These guidelines *and* the introduction of a suitable criterion suggest an acceptable operational definition.

Let us first examine the criterion put forward by Politzer and Parr [17]. These authors offered an approximation for the valence region energy

$$E^v = -\frac{3}{7}(Z - N^c) \int_{r_b}^{\infty} 4\pi r \rho(r)\,dr \tag{1.44}$$

where $\rho(r)$ is the electron density at a distance r from the nucleus. The Politzer–Parr partitioning defines the boundary surface separating the core and valence regions at the minimum of the radial distribution function $R(\mathbf{r}) = R(r) = 4\pi r^2 \rho(r)$. In Hartree–Fock calculations, these minima approximately occur at the 'right places' (in the language of freshmen chemistry), i.e., at $N^c \simeq 2$ e for the first-row and at $N^c \simeq 2$ and $N^c \simeq 10$ e for the second-row elements. Typical examples are given in Figures 1.4 and 1.5. This result casts light on the physical involvement of the electronic shell structure in a meaningful separation of an atom into core and valence regions:

FIGURE 1.4. Radial distribution function of neon.

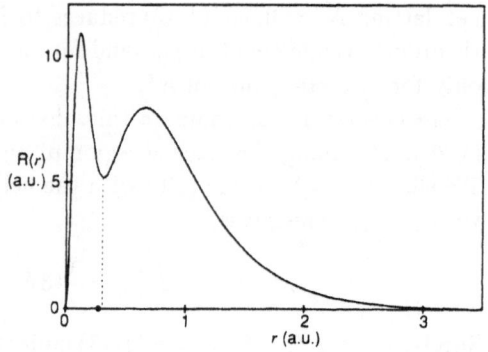

FIGURE 1.5. Radial distribution function of argon.

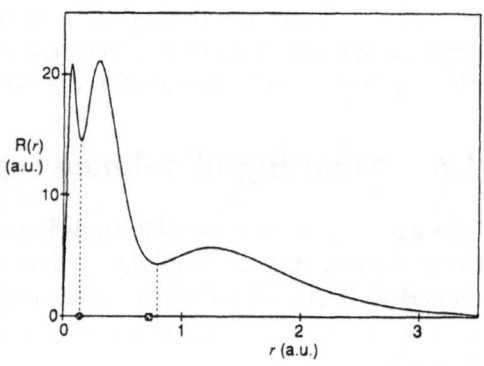

integer core populations are confined within boundaries lying in the vicinity of the minima of the Hartree–Fock radial distribution function. This result is certainly generally correct. As regards more precise numerical details, however, namely as concerns the precise positions of the minima, the results should not be taken too literally with Hartree–Fock wave functions. The latter reproduce quite accurately the electronic shell structure, as revealed by electron diffraction data [18], but small differences evidently persist. A final assessment on this matter is left to studies at a post Hartree–Fock level.

An alternate criterion originates in the properties of exchange integrals. Definite integrals of the form $\int^{\tau_1^v}\int^{\tau_2^c} \ldots d\tau_2\, d\tau_1 + \int^{\tau_2^v}\int^{\tau_1^c} \ldots d\tau_1\, d\tau_2$ describe $1/r_{12}$ interactions between electrons assigned to the core region τ^c and electrons associated with the valence space τ^v (Section 1.2). This concerns both

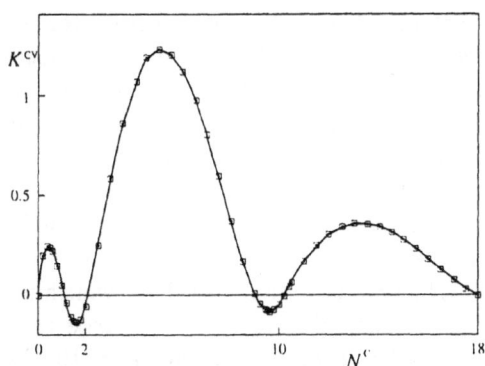

FIGURE 1.6. Argon. Total core-valence exchange, K^{cv}, au

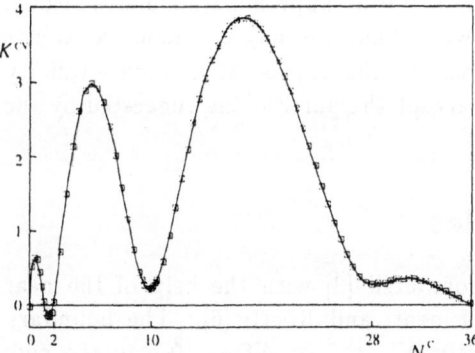

FIGURE 1.7. Krypton. Total core-valence exchange, K^{cv}, au

exchange and Coulomb integrals. Consider the sum, K^{cv}, which collects all the relevant exchange terms between core and valence electrons. This sum can vanish, i.e., $K^{cv} = 0$ is possible[4]. Nonzero exchange integrals between individual electrons are a consequence of their indistinguishability. It seems natural to argue that a group of electrons should not be distinguished from an other group of electrons if the total exchange between these groups is nonzero and that a vanishing K^{cv} should thus accompany a discrimination between core and valence electrons. The examples given in Figures 1.6 and 1.7 are typical. They duplicate the results obtained from the Politzer–Parr criterion. For the first-row elements, the K^{cv} integrals vanish near the points

---

[4]The function $1s(1)2s(1)1s(2)2s(2)$, for example, can be positive or negative depending on whether r_1 and r_2 are on the same side or on opposite sides of the nodal surface. The final sign of this contribution thus depends on the locations of the boundary and nodal surfaces. Negative terms can be part of K^{cv}.

corresponding to $N^c \simeq 2$ e. For the second-row elements, these integrals
vanish near the points corresponding to $N^c = 2$ and $N^c = 10$ e. From nickel
onwards, however, no boundary can be detected for $N^c = 28$ e (except in
Kr, perhaps)—an observation which is consistent with a relatively significant
degree of interpenetration, for third-row atoms, between their $3d$, $4s$, or $4p$
electrons and their $3s$ or $3p$ electrons [19].

It is certainly reassuring to find recognizable features suggesting a meaningful selection of boundary surfaces. It is gratifying that both criteria—that
rooted in the familiar atomic shell structure and that based on vanishing
core–valence exchange integrals—lead to the same results. It is true that
the boundaries are not found at exactly $N^c = 2$, or $N^c = 2$ and $N^c = 10$ e in
Hartree–Fock calculations. If they were, a valid criticism would remind us
of the fact that the Hartree–Fock method is an approximation and we would
still be frustrated of an exact answer. Thus we may conclude: as things
are, no choice is left other than to assume the at least approximate validity
of the Hartree–Fock results and to accept the partitioning suggested by the
criteria advocated in this Section.

1.6   Numerical examples

The results presented here were computed [3] with the help of the near
Hartree–Fock wave functions of Clementi and Roetti [6]. The boundary
surfaces defined by r_b collect exactly $N^c = 2$ or $N^c = 10$ e in the core
regions of the first- and second-row atoms and ions, respectively.

The breakdown of the core populations of the first-row elements is given
in Table 1.1. These examples illustrate the mix-up of electrons that can
be found in two-electron cores. The corresponding valence populations are
readily deduced by difference. These populations are instrumental in the
evaluation of $\sum_i N_i^v \epsilon_i$, for use in eq. (1.41)[5]. (Similar information is given
in [3] for the ten-electron cores of the second-row atoms and ions.)

Table 1.2 reports the breakdown of the kinetic and potential energies for
the atoms Li–Ar. These results permit the calculation of E^v by means of
eq. (1.42).

Let us first compare valence region energies given by eqs. (1.41) and
(1.42) with the appropriate sums of ionization potentials (IP) reported by
Moore [8]. (With the wave functions used here, which are near the Hartree–
Fock limit, eqs. (1.41) and (1.42) give practically the same results.) They are

---

[5]The orbital energies are found in [6].

TABLE 1.1. Core populations of first-row atoms
and ions, calculated for $N^c = 2$ e, au

Atom/Ion	r_b	N^c_{1s}	N^c_{2s}	N^c_{2p}
Li	1.5321	1.96689	0.03311	
Be$^+$	1.0249	1.95081	0.04919	
Be	0.9852	1.93972	0.06028	
B$^+$	0.7323	1.92322	0.07678	
B	0.7019	1.90507	0.05567	0.03926
B$^-$	0.7019	1.90489	0.04975	0.04535
C$^+$	0.5476	1.88114	0.06328	0.05558
C	0.5371	1.87008	0.05249	0.07743
C$^-$	0.5370	1.86967	0.04743	0.08290
N$^+$	0.4365	1.84512	0.05825	0.09663
N	0.4319	1.83795	0.05111	0.11094
N$^-$	0.4331	1.83963	0.04817	0.11220
O$^+$	0.3617	1.81390	0.05605	0.13005
O	0.3603	1.81070	0.05135	0.13795
O$^-$	0.3608	1.81172	0.04867	0.13961
F$^+$	0.3086	1.78802	0.05552	0.15646
F	0.3078	1.78543	0.05163	0.16294
F$^-$	0.3081	1.78611	0.04914	0.16475
Ne$^+$	0.2683	1.76416	0.05531	0.18053
Ne	0.2678	1.76216	0.05197	0.16587

presented in Table 1.3 . The results for the first-row atoms and ions are self-explanatory. Those of the second-row elements reveal discrepancies between calculated and observed quantities, particularly for the larger atoms. Now, the present results were entirely obtained from Hartree–Fock wave functions which, of course, do not reproduce the experimental energies of the atoms. So it is only natural that non-inclusion of correlation and relativistic effects will show up in this type of comparisons. Taking these limitations into account, it seems fair to conclude that the verification of eqs. (1.41) and (1.42) is acceptable both for the neutral atoms and their ions.

The energy of the ion left behind upon removal of the valence electrons (e.g., C^{+4} from C, C$^+$ or C$^-$) also follows from the ground-state properties of its parent atom or ion. Combining eqs. (1.29), (1.30) and (1.37)–(1.40) we have

$$E^{\text{ion}} = \frac{1}{3} \left(V^c_{ne} - V^{cv}_{ee} + \sum_i N^c_i \epsilon_i \right) \qquad (1.45)$$

TABLE 1.2. Kinetic and potential energies of the first-row atoms ($N^c = 2$ e) and the second-row atoms ($N^c = 10$ e), atomic units

Atom	V_{ee}^{cc}	V_{ee}^{vv}	V_{ee}^{cv}	V_{ne}^{c}	V_{ne}^{v}	T^c	T^v
Li	1.6950	0.0083	0.5777	−16.2424	−0.9043	7.3945	0.0384
Be	2.4027	0.3608	1.7257	−30.0606	−3.5742	14.3310	0.2419
B	3.1451	1.2774	3.4167	−48.0162	−8.8811	23.6095	0.9195
C	3.9251	3.1026	5.7319	−70.2060	−17.9300	35.3732	2.3147
N	4.7359	6.1438	8.6658	−96.6656	−31.6866	49.6872	4.7140
O	5.5682	10.7607	12.1270	−127.4428	−50.6335	66.6064	8.2038
F	6.4246	17.2092	16.2147	−162.5366	−76.1308	86.1724	13.2373
Ne	7.3027	25.8012	20.9343	−201.9740	−109.1578	108.4289	20.1172
Na	63.3826	0.0173	2.6176	−386.7950	−2.9388	161.8618	−0.0046
Mg	72.7127	0.3144	6.7913	−470.7413	−8.3033	199.5414	0.0696
Al	82.0525	0.9337	11.7616	−562.9527	−15.5450	241.5141	0.3588
Si	91.4834	2.0718	18.1519	−663.6071	−25.8012	287.9394	0.9078
P	101.0226	3.8696	25.8926	−772.8182	−39.3969	338.9369	1.7765
S	111.1450	6.6401	35.0954	−891.5750	−56.9485	394.5514	2.9470
Cl	120.3905	10.1317	44.8606	−1017.0808	−77.2505	454.8965	4.5689
Ar	130.2521	14.8295	56.3398	−1152.3816	−102.6660	520.1114	6.7003

as well as, using (1.21),

$$E^{ion} = \frac{1}{3}\left[T^c + 2(V_{ne}^c + V_{ee}^{cc})\right]. \tag{1.46}$$

The results obtained from (1.46) and the data collected in Table 1.2 are indicated in Table 1.4. These energies are compared with direct Hartree-Fock–Roothaan computations [6] of the helium isoelectronic series (for the first-row elements) and of the neon isoelectronic series (for the second row). Moreover, a comparison is offered with the corresponding sums of ionization potentials. The overall agreement between these sets of data is generally satisfactory. The sum of the errors made in our calculation of E^{ion} and E^v equals, of course, the genuine error due to the neglect of correlation and relativistic effects.

As a final verification, we examine the valence energy E_{unp}^v defined by eq. (1.26) applying now eq. (1.43) with the help of 'experimental' E^v's (as given by the appropriate sums of ionization potentials). The values of T^v, taken from Table 1.2, represent only a minor part of the corresponding total kinetic energies. Hence we can reasonably anticipate that no serious bias is introduced by their use.

TABLE 1.3. The calculation of valence region energies of first- and second-row atoms and ions, atomic units

Atom/Ion	$V_{ne}^v + V_{ee}^{cv}$	$\sum_i N_i^v \epsilon_i$	E^v	$-\sum_i \mathrm{IP}$
Li	−0.3266	−0.2719	−0.199	−0.198
Be$^+$	−1.0502	−0.8861	−0.645	−0.669
Be	−1.8485	−0.8852	−0.911	−1.012
B$^+$	−3.9546	−2.3090	−2.088	−2.318
B	−5.4643	−1.9901	−2.485	−2.623
B$^-$	−6.2856	−1.2302	−2.505	−2.634
C$^+$	−9.8639	−4.5032	−4.789	−5.026
C	−12.1982	−3.6788	−5.292	−5.440
C$^-$	−13.6919	−2.3800	−5.357	−5.486
N$^+$	−19.6881	−7.4748	−9.054	−9.276
N	−23.0208	−6.0148	−9.679	−9.810
N$^-$	−24.9552	−3.9412	−9.632	−9.805
O$^+$	−34.4029	−11.2814	−15.228	−15.416
O	−38.5065	−8.7778	−15.761	−15.916
O$^-$	−41.3413	−6.0181	−15.786	−15.970
F$^+$	−54.5590	−15.5717	−23.377	−23.572
F	−59.9162	−12.2559	−24.057	−24.212
F$^-$	−63.8081	−8.6761	−24.161	−24.339
Ne$^+$	−81.4482	−20.7634	−34.071	−34.252
Ne	−88.2235	−16.4997	−34.908	−35.045
Na	−0.3211	−0.2909	−0.204	−0.189
Na$^-$	−0.4642	−0.1336	−0.199	−0.218
Mg$^+$	−0.8594	−0.7891	−0.550	−0.553
Mg	−1.5121	−0.8128	−0.775	−0.834
Al$^+$	−2.8818	−1.8535	−1.578	−1.737
Al	−3.7834	−1.5565	−1.780	−1.957
Al$^-$	−4.3554	−1.0048	−1.788	−1.976
Si$^+$	−6.2523	−3.1726	−3.142	−3.490
Si	−7.6493	−2.5985	−3.416	−3.790
Si$^-$	−8.6748	−1.6993	−3.458	−3.841
P$^+$	−11.6578	−4.9103	−5.523	−6.112
P	−13.5044	−3.9887	−5.831	−6.497
P$^-$	−14.8851	−2.5922	−5.826	−6.526
S$^+$	−19.0820	−7.1071	−8.730	−9.783
S	−21.8531	−5.7216	−9.192	−10.163
S$^-$	−23.5378	−3.7605	−9.099	−10.239
Cl$^+$	−29.1644	−9.5514	−12.905	−14.550
Cl	−32.3899	−7.5676	−13.319	−15.026
Cl$^-$	−35.0160	−5.2338	−13.417	−15.159
Ar$^+$	−42.2796	−12.5041	−18.261	−20.653
Ar	−46.3263	−9.9664	−18.764	−21.232

TABLE 1.4. Energies of the two-electron
ions (Li–Ne) and of the ten-electron ions
(Na–Ar), atomic units

Atom	Ref. [6]	Eq. (1.46)	$-\sum^{\text{core}}\text{IP}$
Li	−7.236	−7.233	−7.280
Be	−13.611	−13.662	−13.657
B	−21.986	−22.044	−22.035
C	−32.361	−32.396	−32.416
N	−44.736	−44.724	−44.802
O	−59.111	−59.048	−59.194
F	−75.486	−75.351	−75.595
Ne	−93.861	−93.638	−94.006
Na	−161.677	−161.654	−162.240
Mg	−198.831	−198.839	−199.477
Al	−240.000	−240.095	−240.756
Si	−285.181	−285.436	−286.079
P	−334.370	−334.885	−335.451
S	−387.565	−388.123	−388.874
Cl	−444.764	−446.161	−446.356
Ar	−505.968	−508.049	−507.882

TABLE 1.5. $E^{\text{v}}_{\text{unp}}$ for $N^c = 2$ e
(Li–Ne) and $N^c = 10$ e
(Na–Ar), au

Atom	$E^{\text{v}}_{\text{unp}}$	$-\sum_i N^{\text{v}}_i(\text{IP})$
Li	−0.278	−0.284
Be	−1.397	−1.312
B	−3.475	−3.439
C	−7.003	−7.118
N	−12.358	−12.721
O	−19.772	−20.406
F	−29.699	−30.698
Ne	−42.509	−43.961
Na	−0.286	−0.359
Mg	−1.216	−1.334
Al	−2.756	−2.894
Si	−5.231	−5.364
P	−8.857	−8.932
S	−13.771	−13.931
Cl	−20.255	−19.885
Ar	−28.498	−27.679

The results thus obtained for E^v_{unp} should represent the closest possible approximation to their 'true' values. Surely, these valence energies are for valence regions containing fractional electron populations, those given by eq. (1.15). For comparison we offer the sum of ionization potentials accounting for the fractional populations by linear interpolation[6]. This, of course, represents only a rough estimate, to be taken with a grain of salt, but gives a physical feeling for the energy accumulated in the valence region of an unperturbed atom. Table 1.5 reveals the general agreement between these estimates of E^v_{unp} and supports the basic tenets underlying their derivation.

The general validity of the core–valence partitioning described so far, including the definition of physically meaningful boundaries and the formula given for E^v, is clearly supported by Hartree–Fock–Roothaan computations and pertinent comparisons with experimental results. Unfortunately, the expression given for E^v is not best suited for applications to molecules. We can do better. But this requires better than Hartree–Fock wave functions.

1.7 Configuration interaction calculations

In the Hartree–Fock approximation, the wave function of an n-electron atom that satisfies the antisymmetry requirement is a normalized Slater determinant D,

$$D = \frac{1}{\sqrt{n!}} \begin{vmatrix} \phi_1(1) & \phi_2(1) & \cdots & \phi_n(1) \\ \phi_1(2) & \phi_2(2) & \cdots & \phi_n(2) \\ \vdots & \vdots & & \vdots \\ \phi_1(n) & \phi_2(n) & \cdots & \phi_n(n) \end{vmatrix} \tag{1.47}$$

where $\phi_1, \phi_2, \ldots, \phi_n$ are the orthonormal Hartree–Fock spin-orbitals of that atom.

Configuration interaction (CI) is conceptually the simplest procedure for improving upon the Hatree–Fock approximation. Consider the determinant formed from the n lowest-energy occupied spin-orbitals: this determinant is $|\Psi_0\rangle$ and stands for the appropriate SCF reference state. In addition, consider the determinants formed by promoting one electron from an orbital k to an orbital v which is unoccupied in $|\Psi_0\rangle$: these are the singly excited

---

[6]For the interpolations using fractional electron populations, the second-last IP was included with the $1s$ population in the valence region. Similarly, the $2s$ and $2p$ contributions were subtracted from the valence region by including the corresponding fractions of the larger IP's in the core. The same approximation was used for $3s$ and $3p$ electrons.

determinants $|\Psi_k^v\rangle$. Similarly, consider doubly excited $(k, l \to v, t)$ determinants $|\Psi_{kl}^{vt}\rangle$, etc., up to n-tuply excited determinants. Then use these many-electron wave functions in an expansion describing the CI many-electron wave function $|\Phi_0\rangle$, i.e.,

$$|\Phi_0\rangle = c_0|\Psi_0\rangle + \sum_{k,v} c_k^v|\Psi_k^v\rangle + \sum_{k \geq l, v \geq t} c_{kl}^{vt}|\Psi_{kl}^{vt}\rangle + \cdots \qquad (1.48)$$

(The summation indices prevent double-counting of excited configurations.) Equation (1.48) is a linear variation function. The expansion coefficients c_0, c_k^v, c_{kl}^{vt}, etc., are varied to minimize the variational integral. $|\Phi_0\rangle$ is a better approximation than $|\Psi_0\rangle$. In principle, i.e., if the basis were complete, CI provides an exact solution. Here we use a truncated expansion retaining only determinants D' which differ from $|\Psi_0\rangle$ by at most two spin-orbitals: this is a singly and doubly excited CI (SDCI).

The presence of excited determinants in $|\Phi_0\rangle$ introduces integrals of the type $\int D' \hat{A} D \, d\tau$, where $\hat{A}$ is an operator. Following the Condon–Slater rules, these n-electron integrals can be reduced to sums of one- and two-electron integrals [20]. Consider two determinants D and D', written as in (1.47), arranged so as to have as many as possible of their left-hand columns match. A one-electron operator $\hat{f}_i$ (namely $-\frac{1}{2}\nabla_i^2$ or $-Z/r_i$) introduces the new integral

$$\int \phi_n'(1)\hat{f}_1\phi_n(1)\,d\tau_1 \qquad (1.49)$$

when D and D' differ by one spin-orbital, $\phi_n' \neq \phi_n$. No contribution arises when D and D' differ by two or more spin-orbitals. The two-electron operator $|1/r_{12}|$ introduces the following new integrals

$$\sum_{j=1}^{n-1} \left[\iint \phi_n'(1)\phi_j(2) \frac{1}{r_{12}} \phi_n(1)\phi_j(2)\,d\tau_1\,d\tau_2 \right.$$

$$\left. - \iint \phi_n'(1)\phi_j(2) \frac{1}{r_{12}} \phi_j(1)\phi_n(2)\,d\tau_1\,d\tau_2 \right] \qquad (1.50)$$

when D and D' differ only by one spin-orbital, $\phi_n' \neq \phi_n$, or

$$\iint \phi_n'(1)\phi_{n-1}'(2) \frac{1}{r_{12}} \phi_n(1)\phi_{n-1}(2)\,d\tau_1\,d\tau_2$$

$$- \iint \phi_n'(1)\phi_{n-1}'(2) \frac{1}{r_{12}} \phi_{n-1}(1)\phi_n(2)\,d\tau_1\,d\tau_2 \qquad (1.51)$$

when D and D' differ by two spin-orbitals, i.e., $\phi'_n \neq \phi_n$ and $\phi'_{n-1} \neq \phi_{n-1}$. No contribution arises when D and D' differ by three or more spin-orbitals. The integrals (1.49)–(1.51) involve summation over the appropriate spin co-ordinates and integration over spatial coordinates. Inspection of these integrals indicates that they are amenable to a space partitioning along the lines described in Section 1.2, i.e., simply by selecting the appropriate limits of integration. The arguments developed with eqs. (1.17)–(1.19) still hold, although they must be reinterpreted in light of the new functions found in (1.49)–(1.51). Briefly, we can now approach the study of core and valence regions with the help of CI wave functions.

Exploratory calculations reveal relatively small differences between SDCI and SCF results (Table 1.6) [21]. As one would expect, improvements are achieved for the energies but, as usual in this type of work where the rate of convergence is rather slow, it is obvious that many more configurations are required to approach accurate non-relativistic solutions. Temporarily we cannot offer clear-cut answers to questions like: 'what are the numbers of core electrons at the exact minima of the radial distribution function?' or 'how close to 0 are the core-valence integrals for $N^c = 2$ or 10 electrons for first- and second-row atoms, respectively?' Final statements about these (and related) questions must await final calculations which should preferably include relativistic corrections as well.

TABLE 1.6. Carbon atom. Selected SCF and SDCI results obtained for $N^c = 2$ electron, atomic units

Property	4-31G		6-31G	
	SCF	SDCI	SCF	SDCI
E	−37.5446	−37.6056	−37.5882	−37.6490
V^v_{ne}	−17.8586	−17.9148	−17.7928	−17.8355
V^c_{ne}	−70.0880	−70.0585	−70.1804	−70.1493
V^{vv}_{ee}	3.1917	3.1859	3.1751	3.1664
V^{cc}_{ee}	3.9161	3.8936	3.9143	3.8906
V^{cv}_{ee}	5.7168	5.7102	5.6983	5.6888
T^v	2.3901	2.4143	2.2285	2.2449
T^c	35.1870	35.1637	35.3688	35.3451
E^v	−5.1700	−5.2077	−5.2034	−5.2386
E^{ion}	−32.3856	−32.3887	−32.3878	−32.3908
K^{cv}	−0.0356	−0.0320	−0.0357	−0.0321
$-V/T$	1.9991	2.0007	1.9998	2.0016

On the other hand, these CI explorations—however crude they may be—
solidify the view that one can reasonably proceed with the ideas developed
so far from inspection of accurate SCF results as they surely represent valid
approximations. It must be added that the calculation of accurate atomic
valence-region energies is not our primary goal. Here we are first and fore-
most interested in the formal description of our core–valence separation so
that we may adapt it to molecules. With this idea in mind, we shall now
develop a handy formula for the valence-region energy of atoms.

1.8   Alternate energy formula

Consider eq. (1.39) and examine the physical nature of its V_{ee}^{cv} part. This
potential energy describes the repulsion between an outer electron cloud (the
valence electrons) and the N^c core electrons. Gauss' theorem tells us that
the V_{ee}^{cv} potential is just as though all the core electronic charge were lumped
at the nuclear position. Hence, the $\int_{r_b}^{\infty}[\rho(r)/r]\mathrm{d}r$ integrals being the same
in the evaluation of V_{ee}^{cv} and V_{ne}^{v}, and taking into account that the nuclear
charge Z has been replaced by N^c, it follows that

$$V_{ee}^{cv} = -\frac{N^c}{Z}V_{ne}^{v}. \tag{1.52}$$

It is now clear that

$$V_{ne}^{v} + V_{ee}^{cv} = -(Z - N^c)\int_{r_b}^{\infty}\frac{\rho(r)}{r}\mathrm{d}r \tag{1.53}$$

represents the nuclear–electronic potential energy of the outer electrons in
the field of an expanded 'effective nucleus' $(Z - N^c)$. This is an approximation, of course, because it does not consider the spin of the core electrons.
Now we know that the exchange part, K^{cv}, of the core–valence interactions
is certainly very small for the r_b that defines the boundary between core and
valence regions. These results, which are in general agreement with those
of Politzer and Daiker [19], suggest that (1.53), though approximate, is not
bad at all. Note that because of (1.52) we can also write

$$V_{ne}^{v} + V_{ee}^{cv} = \frac{Z - N^c}{Z}V_{ne}^{v}. \tag{1.54}$$

Our simple use of Gauss' theorem for the evaluation of V_{ee}^{cv} does not cause
any significant bias in applications of eq. (1.39) [3]. In defence of eqs. (1.53)

and (1.54) one could possibly add that monoelectronic properties are better reproduced in Hartree–Fock theory than bielectronic properties.

The simplification introduced with the use of Gauss' theorem is valuable for the physical picture it conveys, that of a valence electron cloud in the field of a nucleus partially screened by its core electrons. Using it in (1.39) we get

$$E^{\mathrm{v}} = -\frac{1}{\gamma^{\mathrm{v}}}(Z - N^{\mathrm{c}}) \int_{r_{\mathrm{b}}}^{\infty} \frac{\rho(r)}{r} \mathrm{d}\mathbf{r}. \qquad (1.55)$$

Now we know that, except for Li, Be, Na, and Mg, γ^{v} is close to 7/3 [3]. Remembering that, given spherical symmetry, it is $\mathrm{d}\mathbf{r} = 4\pi r^2 \mathrm{d}r$, we see that eq. (1.55) is our counterpart of the Politzer–Parr Thomas–Fermi-like formula (1.44) describing the valence region energy of atoms. Here we should stress that *equation* (1.55) *represents a valence energy in which relaxation effects are included.* It was derived from eq. (1.39) and cannot be mistaken for $E^{\mathrm{v}}_{\mathrm{unp}}$—an important aspect which was not recognizable in earlier derivations of energy formulas like eq. (1.55) in the spirit of Thomas–Fermi theory.

This derivation of the energy formula for atomic valence-region electrons foreshadows a similar one that shall be worked out for molecules.

1.9   Summary

The mental subdivision of a Hartree–Fock atom into an inner core and an outer valence region leads to simple statistics regarding the eigenvalues, ϵ_i, of the spin orbitals. The electronic energies statistically accumulated in the core and valence regions are adequately described by $\sum_i N_i^{\mathrm{c}} \epsilon_i$ and $\sum_i N_i^{\mathrm{v}} \epsilon_i$, respectively, where N_i^{c} and N_i^{v} are the corresponding electron populations. The radius r_{b} which defines the boundary between the core and valence regions determines the values of N_i^{c} and N_i^{v}. For any r_{b} of our choice we have $\sum_i N_i^{\mathrm{v}} \epsilon_i = T^{\mathrm{v}} + V_{\mathrm{ne}}^{\mathrm{v}} + 2V_{\mathrm{ee}}^{\mathrm{vv}} + V_{\mathrm{ee}}^{\mathrm{cv}}$ (with a similar expression for the core region), where each quantity is a function of r_{b}. The difference between the left- and right-hand sides of this equation is everywhere 0 for true Hartree–Fock wave functions, but not so for approximations thereof. This difference, and its profile with varying r_{b}'s, can be used as a quality test indicating how and in which regions approximate expansions $\phi_i = \sum_k c_{ki} \chi_k^{lm}$ differ from genuine Hartree–Fock eigenfunctions. This way of assessing the merits of approximate wave functions in different regions of space presents conceptual similarities with a criterion advocated by Javor *et al.* [22] and with King's concept of 'reduced local energy' as a criterion for the accuracy of Hartree–

Fock wave functions [23]. All these tests avoid the pitfalls and limitations of the common 'global tests' where inaccuracies in the wave function in one region of configuration space may offset inaccuracies in another region of configuration space, with the result that computed expectation values may be fortuitously in good agreement with experimental values.

The key to meaningful—i.e., observable—valence-region energies E^v is rooted in Hartree–Fock identities, namely, i) in a relationship between the total energy of the atom and the sum of its occupied orbital energies, on one hand, and ii) between the total energy and the nuclear–electronic potential energy, on the other. A combination of these relationships leads to a simple formula for E^v, the energy that mirrors the actual removal of electrons from the valence shell, i.e., $E^v = \frac{1}{3}(T^v + 2V^v)$, where V^v is the total potential energy of the $\sum_i N_i^v$ electrons associated with the outer (valence) region. The companion formula for the ion left behind upon removal of the valence electrons is $E^{ion} = \frac{1}{3}[T^c + 2(V_{ne}^c + V_{ee}^{cc})]$. Both E^v and E^{ion} are meaningful only for discrete values of N^c, namely integer core populations confined within boundaries lying at (or close to) the minima of the radial distribution function. This result casts light on the physical involvement of the electronic shell structure in a meaningful separation of an atom into core and valence regions—a separation also characterized by vanishing core–valence exchange integrals. We find, as expected, $N^c = 2$ or 10 electrons for first- and second-row elements, respectively.

Finally, a considerable simplification is introduced with the use of Gauss' theorem. The valence-region energy is now readily described in terms of the potential of a valence electron-cloud in the field of an expanded, effective nucleus screened by its core electrons. This form, reminiscent of what can be deduced from Thomas–Fermi theory, proves most useful in our forthcoming applications to molecular systems.

Bibliography

[1] H. Eyring, J. Walter, and G. E. Kimball,"Quantum Chemistry", John Wiley & Sons, Inc., New York, 1954.

[2] N. Desmarais, G. Dancausse, and S. Fliszár, *Can. J. Chem.*, **71**, 175 (1993).

[3] S. Fliszár, N. Desmarais, and G. Dancausse, *Can. J. Chem.*, **70**, 537 (1992).

[4] D. R. Hartree, "The Calculation of Atomic Structures", Wiley, New York, 1957; C. Froese Fischer, "The Hartree–Fock Method for Atoms: a Numerical Approach", Wiley, New York, 1977.

[5] J. B. Mann, "Atomic Structure Calculations", Los Alamos Scientific Laboratory, University of California, Los Alamos, NM, Part I: Hartree–Fock Energy Results for the Elements Hydrogen to Lawrencium, 1967. Part II: Hartree–Fock Wave Functions and Radial Expectation Values, 1968.

[6] E. Clementi and C. Roetti, *At. Data Nucl. Data Tables*, **14**, 179 (1974).

[7] W. J. Hehre, L. Radom, P. v.R. Schleyer, and J. A. Pople, "Ab Initio Molecular Orbital Theory", Wiley, New York, 1986.

[8] C. E. Moore, *Natl. Stand. Ref. Data Ser.* (U.S. Natl. Bur. Stand.), **34** (1970).

[9] R. N. Dixon and I. L. Robertson *in* Specialist Periodical Reports, Theoretical Chemistry, vol. 3, The Chemical Society, London, 1978, Chapter 4.

[10] H. Hellmann, "Einführung in die Quantenchemie", F. Deuticke and Co., Leipzig, 1937; R. P. Feynman, *Phys. Rev.*, **56**, 340 (1939).

[11] N. H. March, *Adv. Phys.*, **6**, 1 (1957).

[12] E. A. Milne, *Proc. Cambridge Philos. Soc.*, **23**, 794 (1927).

[13] J. Goodisman, *Phys. Rev.*, **A2**, 1193 (1970); *Theor. Chim. Acta*, **24**, 1 (1972).

[14] R. G. Parr, S. R. Gadre, and L. J. Bartolotti, *Proc. Natl. Acad. Sci., U.S.A.*, **76**, 2522 (1979).

[15] K. Ruedenberg, *J. Chem. Phys.*, **66**, 375 (1977).

[16] P. Politzer, *J. Chem. Phys.*, **64**, 4239 (1976); **69**, 491 (1978).

[17] P. Politzer and R. G. Parr, *J. Chem. Phys.*, **64**,4634 (1976).

[18] L. S. Bartell and L. O. Brockway, *Phys. Rev.*, **90**, 833 (1953).

[19] P. Politzer and K. C. Daiker, *Chem. Phys. Lett.*, **20**, 309 (1973).

[20] R. G. Parr, "Quantum Theory of Molecular Electronic Structure", Benjamin, New York, 1963.

[21] N. Desmarais and S. Fliszár, to be published.

[22] F. Javor, G. F. Thomas, and S. M. Rothstein, *Int. J. Quantum Chem.*, **11**, 59 (1977).

[23] F. W. King, M. K. Kelly, and M. A. LeGore, *J. Chem. Phys.*, **76**, 574 (1982); F. W. King, M. A. LeGore, and M. K. Kelly, *J. Chem. Phys.*, **75**, 809 (1981).

Chapter 2

The Valence Region of Molecules

2.1 Scope

This Chapter is about molecules. A molecule is a collection of nuclei Z_k, $Z_l, \ldots$ at rest (in the Born–Oppenheimer approximation) in a sea of fast-moving electrons. The nuclei can be identified and thus provide convenient reference marks. Each nucleus found in a molecule, say, Z_k, can be viewed as the nucleus of a 'giant atom' extending over the entire molecule: Z_k is surrounded by charges, just like in an ordinary atom, with the difference that motionless positive point charges are now part of its environment. Of course, the electronic content is still described by its stationary density.

This abridged description hints at the strategy adopted in our calculations but requires clarification. When an atom becomes part of a molecule, it is clear that both its nucleus *and* its electrons are incorporated. So, when we say that Z_k is in a molecule we refer more precisely to what we shall call 'atom k in the molecule' meaning i) that Z_k has entered the molecule together with its electrons—those of the isolated atom k— and ii) that the incorporated atom k differs from the isolated atom k. This vision of an atom in a molecule does not introduce any physical constraint. The molecule is still taken as a whole, as described above. No spatial partitioning of the molecule is considered, carrying the picture of subspaces defined by boundaries enclosing one nucleus and a share of the electronic charge entirely distributed among the subspaces. The only charge partitioning contemplated here concerns a separation into core and valence regions along the lines described in

Chapter 1 for the isolated atoms. This separation lays the foundations of the theory describing bond energies.

The argument is developed in three steps: i) the bare nucleus, Z_k, of atom k in the molecule is considered in the field of all the electrons and of all the other nuclei found in the molecule. This introduces the notion of binding. ii) Z_k is considered together with its core electrons, N_k^c. This amounts to a core–valence partitioning of 'atom k in the molecule': its valence region now consists of all the electrons but N_k^c and of all the other nuclei. At this stage, it is not taken into account that the other nuclei—those embedded in the valence region of atom k—may possess core electrons of their own. The final step iii) considers all the nuclei with their core electrons, meaning that all the remaining electrons shall be regarded as the valence electrons of the molecule. The idea is to find an expression for the energy of a molecule featuring the role of its electronic valence region.

2.2   Basic formulas

Before getting started, let us review some topics of basic theory and present well-known formulas with intent to define a string of useful terms.

Consider the total energy

$$E^{\text{mol}} = \langle \Psi^{\text{mol}} | \hat{H}^{\text{mol}} | \Psi^{\text{mol}} \rangle \tag{2.1}$$

of a molecule at its potential minimum. The Hamiltonian in (2.1) is for an n-electron molecule in the Born–Oppenheimer approximation, i.e., in atomic units,

$$\hat{H}^{\text{mol}} = -\frac{1}{2}\sum_i \nabla_i^2 - \sum_k \sum_i \frac{Z_k}{r_{ik}} + \sum_j \sum_{i>j} \frac{1}{r_{ij}} + \sum_k \sum_{l>k} \frac{Z_k Z_l}{R_{kl}} \tag{2.2}$$

where i and j refer to electrons and k and l refer to nuclei. The first term is the operator for the kinetic energy of the electrons. The second term represents the attraction between the nuclei and the electrons, r_{ik} being the distance between electron i and nucleus k. The third term represents the repulsions between electrons i and j at a distance r_{ij}. The last term represents the repulsions between the nuclei, R_{kl} being the distance between nuclei with charges Z_k and Z_l.

We want to calculate the derivative $(\partial E^{\text{mol}}/\partial Z_k)_\rho$ at constant electron density ρ with respect to the nuclear charge of one of its nuclei, treated as a continuous variable. The Hellmann–Feynman theorem and inspection of the Hamiltonian (2.2) indicate that

$$\left(\frac{\partial E^{\mathrm{mol}}}{\partial Z_k}\right)_\rho = \left\langle \Psi^{\mathrm{mol}} \left| \frac{\partial \hat{H}^{\mathrm{mol}}}{\partial Z_k} \right| \Psi^{\mathrm{mol}} \right\rangle$$

$$= -\int \frac{\rho(\mathbf{r})}{|\mathbf{r} - \mathbf{R}_k|}\,d\mathbf{r} + \sum_{l \neq k} \frac{Z_l}{R_{kl}} \tag{2.3}$$

where $\rho(\mathbf{r})$ is the electron density in the volume element $d\mathbf{r}$ at the point $\mathbf{r}$ and $\mathbf{R}_k$ defines the position of nucleus Z_k. The derivative $(\partial E^{\mathrm{mol}}/\partial Z_k)_\rho$ represents the total potential at the site of Z_k. The potential energy that involves Z_k is thus

$$V_k = Z_k \left(\frac{\partial E^{\mathrm{mol}}}{\partial Z_k}\right)_\rho \tag{2.4}$$

i.e., using (2.3),

$$V_k = -Z_k \int \frac{\rho(\mathbf{r})}{|\mathbf{r} - \mathbf{R}_k|}\,d\mathbf{r} + Z_k \sum_{l \neq k} \frac{Z_l}{R_{kl}}. \tag{2.5}$$

In eq. (2.5), Z_k 'sees' all the electrons and all the nuclei $Z_{l \neq k}$ of the molecule. The first right-hand side term of (2.5) is the nuclear–electronic potential energy involving Z_k, i.e.,

$$V_{\mathrm{ne},k} = -Z_k \int \frac{\rho(\mathbf{r})}{|\mathbf{r} - \mathbf{R}_k|}\,d\mathbf{r} \tag{2.6}$$

and the second term represents the repulsion between Z_k and all the other nuclei of the molecule. Summation of all the $V_{\mathrm{ne},k}$ terms over all the nuclei gives the total nuclear–electronic potential energy of the molecule, namely

$$V_{\mathrm{ne}} = -\sum_k Z_k \int \frac{\rho(\mathbf{r})}{|\mathbf{r} - \mathbf{R}_k|}\,d\mathbf{r}. \tag{2.7}$$

Equation (2.5) shows that the sum $\sum_k V_k$ involves $\sum_k Z_k(\sum_{l \neq k} Z_l/R_{kl})$. This introduces a double counting of the total internuclear repulsion

$$V_{\mathrm{nn}} = \sum_k \sum_{l > k} \frac{Z_k Z_l}{R_{kl}}. \tag{2.8}$$

It is therefore, for the entire molecule,

$$\sum_k V_k = V_{\mathrm{ne}} + 2V_{\mathrm{nn}}. \tag{2.9}$$

2.3   The atom in a molecule. Binding

The molecule is considered in its equilibrium geometry, meaning that we can apply the molecular virial theorem, $2E^{\text{mol}} = V_{\text{ne}} + V_{\text{ee}} + V_{\text{nn}}$, where V_{ee} is the total interelectronic repulsion energy. Using (2.4) and (2.9) we get

$$2E^{\text{mol}} - \sum_k Z_k \left(\frac{\partial E^{\text{mol}}}{\partial Z_k} \right)_\rho = V_{\text{ee}} - V_{\text{nn}}. \qquad (2.10)$$

Hartree–Fock wave functions satisfy the virial theorem [1]. The Hellmann–Feynman theorem is obeyed by Hartree–Fock (as well as exact) wave functions [2]. Briefly, (2.10) is valid in Hartree–Fock theory. Taking the internuclear repulsion (2.8) into account, the Hartree–Fock formula becomes

$$E^{\text{mol}} = \sum_i \nu_i \epsilon_i - (V_{\text{ee}} - V_{\text{nn}}). \qquad (2.11)$$

Combining (2.10) and (2.11) to get rid of the $(V_{\text{ee}} - V_{\text{nn}})$ part, we can write

$$(3 - \gamma)E^{\text{mol}} = \sum_i \nu_i \epsilon_i \qquad (2.12)$$

where γ is defined as

$$\gamma = \frac{1}{E^{\text{mol}}} \sum_k Z_k \left(\frac{\partial E^{\text{mol}}}{\partial Z_k} \right)_\rho. \qquad (2.13)$$

Finally, remembering (2.4) and (2.9), we rewrite (2.13) as follows

$$E^{\text{mol}} = \frac{1}{\gamma}(V_{\text{ne}} + 2V_{\text{nn}}). \qquad (2.14)$$

For isolated atoms, eqs. (2.12)–(2.14) reduce to (1.34)–(1.36). Equations (2.12) and (2.14) represent an identity in Hartree–Fock theory provided, of course, that all quantities are evaluated at equilibrium geometry [3,4]. The parameter γ is close to $\frac{7}{3}$. This limit generates the Politzer formula [5], $E^{\text{mol}} = \frac{3}{7}(V_{\text{ne}} + 2V_{\text{nn}})$, which has been studied in detail [4]. The electronic kinetic energy and the interelectronic repulsion energy have disappeared in (2.14): E^{mol} is expressed in terms of the potentials at the nuclei. This result follows from (2.3). It is important because the individual nuclei can be identified in a molecule. Equation (2.14) is pivotal in the calculations described further below.

At this point we can examine an important property relating to atoms in a molecule. Consider the energy defined in (2.15):

$$E_k = \frac{1}{\gamma_k} V_k. \tag{2.15}$$

The parameter γ_k is an atomic parameter. It carries the label of 'atom k in the molecule', as does V_k. There is a constraint on γ_k, namely: the average of the $(1/\gamma_k)$'s, weighted by V_k, must restore the $1/\gamma$ of (2.13) [4]. In other words, we demand that

$$\frac{1}{\gamma} = \frac{\sum_k (\frac{1}{\gamma_k}) V_k}{\sum_k V_k}. \tag{2.16}$$

This definition of γ_k and eqs. (2.9), (2.14) and (2.15) lead to

$$\sum_k E_k = E^{\text{mol}}. \tag{2.17}$$

What is E_k? It is tempting to call it the energy of an atom in a molecule. We do so for the sake of simplicity, but shall not forget that the subscript "k" associates E_k with an atom that is solely identified by its nucleus. There is one E_k for each nucleus. These E_k's have the desirable property that their sum is the total energy of the molecule. Therein lies their usefulness, unabated by the fact that the individual E_k's are not amenable to direct measurements.

The energy E_k differs, in principle, from the energy of the isolated ground-state atom k, E_k^{atom}. The difference

$$\Delta E_k = E_k^{\text{atom}} - E_k \tag{2.18}$$

contains part of the molecular binding energy. A good insight into the meaning of ΔE_k is offered by the atomization energy, ΔE_a^*, defined as

$$\begin{aligned}
\Delta E_a^* &= \text{energy of all the isolated ground–state atoms} \\
&\quad \text{minus the ground–state energy of the molecule} \\
&= \sum_k E_k^{\text{atom}} - E^{\text{mol}}.
\end{aligned} \tag{2.19}$$

It follows immediately from eqs. (2.17)–(2.19) that

$$\sum_k \Delta E_k = \Delta E_a^*. \tag{2.20}$$

Equation (2.20) offers an atom-by-atom partitioning of the molecular binding energy. The nice thing about (2.20) is that it does not imply any spatial partitioning of the molecule. Equation (2.18) is instrumental in the theory of bond energies. ΔE_a^* is convenient for comparisons with experimental results.

The energy given in (2.15) for E_k refers to a bare nucleus, Z_k, in the field of all the electrons and all the other nuclei. For an isolated atom, (2.15) becomes (1.36). We can now proceed with a core–valence separation along the lines described in Chapter 1.

2.4   Core–valence partitioning of molecules

Let us begin with the potential energy V_k described by eq. (2.5). The part representing the nuclear–electronic interaction contains a definite integral extending over the full range of coordinates. This integral can be decomposed as follows, in short-hand notation,

$$\int \frac{\rho(\mathbf{r})}{|\mathbf{r} - \mathbf{R}_k|}\, d\mathbf{r} = \int_{\mathbf{R}_k}^{r_{b,k}} \dots d\mathbf{r} + \int_{r_{b,k}}^{\infty} \dots d\mathbf{r}. \qquad (2.21)$$

In (2.21) $r_{b,k}$ defines the boundary of the core region associated with nucleus Z_k at the point $\mathbf{R}_k$. This region contains

$$N_k^c = \int_{\mathbf{R}_k}^{r_{b,k}} \rho(\mathbf{r})\, d\mathbf{r}$$

core electrons. For the reasons given in Chapter 1.5, N_k^c is an integer, e.g., $N_k^c = 2$ e if k is a first-row atom. The first right-hand side integral in (2.21) refers to the nuclear–electronic potential energy involving Z_k and N_k^c:

$$V_{ne,k}^c = -Z_k \int_{\mathbf{R}_k}^{r_{b,k}} \frac{\rho(\mathbf{r})}{|\mathbf{r} - \mathbf{R}_k|}\, d\mathbf{r}. \qquad (2.22)$$

The second integral extends over the entire region beyond the boundary surface defined by $r_{b,k}$, briefly, the valence region of Z_k in the molecule. Hence it includes the core electrons $N_l^c, \dots$ associated with the nuclei $Z_l, \dots$, etc. found in the valence region of Z_k. We can thus rewrite (2.5) as follows

$$V_k = \left[-Z_k \int_{r_{b,k}}^{\infty} \frac{\rho(\mathbf{r})}{|\mathbf{r} - \mathbf{R}_k|}\, d\mathbf{r} + Z_k \sum_{l \neq k} \frac{Z_l}{R_{kl}} \right] + V_{ne,k}^c. \qquad (2.23)$$

The term in brackets describes the interaction between Z_k and all the charges found beyond its core region. (For an isolated atom, this term is simply V_{ne}^v, eq. (1.17), and (2.23) becomes $V_{ne} = V_{ne}^v + V_{ne}^c$.)

Now we know that the core electrons associated with Z_k also interact with the charges of the valence region. First, they repel the valence electrons (this is the V_{ee}^{cv} term described for the atoms) and thus reduce the effective attraction exerted by the nuclear charge. In the model adopted here, that of a charged inner sphere of radius r_b surrounded by valence electrons, the nuclear–electronic attraction and this core–valence repulsion play similar physical roles, one interaction opposing the other, and are considered together. Second, the core electrons N_k^c attract the nuclei $Z_l \ldots$ of the valence region of Z_k and thus counteract the repulsion between Z_k and the other nuclei. These repulsions and counteracting attractions are also considered together. In short, the core electrons not only screen the attraction between Z_k and the outer electrons but also screen the internuclear repulsions involving Z_k. The total screening is appropriately described by

$$V_k^{cv} = \text{interaction energy between } N_k^c \text{ and the electronic}$$
$$\text{and nuclear charges found outside the core region} \quad (2.24)$$

Using (2.24) we rewrite (2.23) in the following manner

$$V_k = \left[-Z_k \int_{r_{b,k}}^{\infty} \frac{\rho(\mathbf{r})}{|\mathbf{r} - \mathbf{R}_k|}\, d\mathbf{r} + Z_k \sum_{l \neq k} \frac{Z_l}{R_{kl}} + V_k^{cv} \right] + (V_{ne,k}^c - V_k^{cv}). \quad (2.25)$$

Finally, we obtain E_k from (2.15), i.e.,

$$E_k = \frac{1}{\gamma_k^v} \left[-Z_k \int_{r_{b,k}}^{\infty} \frac{\rho(\mathbf{r})}{|\mathbf{r} - \mathbf{R}_k|}\, d\mathbf{r} + Z_k \sum_{l \neq k} \frac{Z_l}{R_{kl}} + V_k^{cv} \right] + \frac{1}{\gamma_k^c}(V_{ne,k}^c - V_k^{cv}). \quad (2.26)$$

where $1/\gamma_k$ of (2.15) is the average of $1/\gamma_k^v$ (weighted by the term in brackets) and $1/\gamma_k^c$ (with a weight of $V_{ne,k}^c - V_k^{cv}$). This derivation has followed the pattern developed for the atoms (Chapter 1.4), with the difference that the nuclear charges $Z_l \ldots$ found in the valence region are now taken care of. For an atom, eq. (2.25) reduces to $V_{ne} = (V_{ne}^v + V_{ee}^{cv}) + (V_{ne}^c - V_{ee}^{cv})$ and the transformation from (2.25) to (2.26) is the same that gave (1.38), the atomic counterpart of (2.26). The first part of the right-hand side of (2.26) thus represents the valence-region energy of atom k in the molecule and the second term is the energy, E_k^{ion}, of the ion containing Z_k and the appropriate number of core electrons. So we simplify things by writing

$$E_k = \frac{1}{\gamma_k^{\mathrm{v}}} \left[-Z_k \int_{r_{\mathrm{b},k}}^{\infty} \frac{\rho(\mathbf{r})}{|\mathbf{r} - \mathbf{R}_k|} \, d\mathbf{r} + Z_k \sum_{l \neq k} \frac{Z_l}{R_{kl}} + V_k^{\mathrm{cv}} \right] + E_k^{\mathrm{ion}}. \qquad (2.27)$$

The total energy of the molecule is given by eq. (2.17). Hence

$$E^{\mathrm{mol}} = \frac{1}{\gamma^{\mathrm{v}}} \sum_k \left[-Z_k \int_{r_{\mathrm{b},k}}^{\infty} \frac{\rho(\mathbf{r})}{|\mathbf{r} - \mathbf{R}_k|} \, d\mathbf{r} + Z_k \sum_{l \neq k} \frac{Z_l}{R_{kl}} + V_k^{\mathrm{cv}} \right] + \sum_k E_k^{\mathrm{ion}} \quad (2.28)$$

where $1/\gamma^{\mathrm{v}}$ is the appropriate average of the individual $(1/\gamma_k^{\mathrm{v}})$'s, weighted by the terms in brackets in (2.28). For atoms, (2.28) reduces to $E = E^{\mathrm{v}} + E^{\mathrm{ion}}$.

The first right-hand side term of (2.28) describes the valence-region energy of a molecule, $E^{\mathrm{v,mol}}$. (Examine Table 2.1 to gain familiarity with the quantities involved: $E^{\mathrm{v,mol}}$ represents about 20% of E^{mol}.) A more instructive form of (2.28) can be obtained as follows.

2.5   The valence region energy

A considerable simplification can be achieved with the help of Gauss' theorem. This theorem offers a simple solution for the V_k^{cv} term found in (2.28).

Gauss' theorem tells us that if the charge of a subshell is rigorously spherically symmetrical about a nucleus, the effect felt by a point charge 'outside' such a subshell is though all the 'inner' charge were concentrated at the nucleus to form a positively charged core which acts as a point charge.

TABLE 2.1. Valence-region energy of selected
molecules, atomic units

Molecule	E^{mol}	$E^{\mathrm{v,mol}}$	$\sum_k E_k^{\mathrm{ion}}$
Ethane, C_2H_6	-79.8439	-15.0120	-64.8319
Ethene, C_2H_4	-78.6076	-13.7756	-64.8319
Cyclohexane, C_6H_{12}	-235.9408	-41.4450	-194.4958
Benzene, C_6H_6	-232.3125	-37.8166	-194.4958
Acetone, C_3H_6O	-193.2336	-36.7918	-156.4418
1,4-Dioxane, $C_4H_8O_2$	-307.7994	-59.7476	-248.0517

The core–valence interelectronic Coulomb energy would involve the same $\int_{r_b}^{\infty} [\rho(r)/r]\, dr$ integral as the nuclear–electronic potential energy (Chapter 1.7). We cannot be sure how closely Gauss' theorem is obeyed by the core electrons found in molecules. Yet, considering the dominant topological properties of the charge density in molecules, revealed by the local maxima it exhibits at the nuclei and by its nearly spherical decay away from these points [6], and relying on the information gathered for the atoms, it seems safe to assume that Gauss' theorem represents a valid approximation at least for the problem at hand. Electronic charge density contour maps, particularly in the short range containing the appropriate number of core electrons, certainly point in that direction. So we approximate (2.24) as follows[1]

$$V_k^{cv} = N_k^c \int_{r_{b,k}}^{\infty} \frac{\rho(\mathbf{r})}{|\mathbf{r} - \mathbf{R}_k|}\, d\mathbf{r} - N_k^c \sum_{l \neq k} \frac{Z_l}{R_{kl}}. \qquad (2.29)$$

The first right-hand side term of (2.29) describes the repulsion between N_k^c, located at the point $\mathbf{R}_k$, and all the outer electrons. The second term describes the attraction between N_k^c and all the nuclei other than Z_k. Defining now the *effective* nuclear charge Z_k^{eff} as

$$Z_k^{eff} = Z_k - N_k^c \qquad (2.30)$$

we obtain from (2.28) and (2.29) that

$$E^{mol} = \frac{1}{\gamma^v} \sum_k \left[-Z_k^{eff} \int_{r_{b,k}}^{\infty} \frac{\rho(\mathbf{r})}{|\mathbf{r} - \mathbf{R}_k|}\, d\mathbf{r} + Z_k^{eff} \sum_{l \neq k} \frac{Z_l}{R_{kl}} \right] + \sum_k E_k^{ion}. \qquad (2.31)$$

The integral appearing in (2.31) includes all the electrons outside the boundary $r_{b,k}$, hence also the core electrons $N_l^c, \ldots$ of nuclei $Z_l, \ldots$ Consider instead a 'truncated' integral $\int^{val} \ldots d\mathbf{r}$ that avoids the core electrons of all atoms, including N_k^c. In this manner, the systematic use of Gauss' theorem leads to

$$\int_{r_{b,k}}^{\infty} \frac{\rho(\mathbf{r})}{|\mathbf{r} - \mathbf{R}_k|}\, d\mathbf{r} = \int^{val} \frac{\rho(\mathbf{r})}{|\mathbf{r} - \mathbf{R}_k|}\, d\mathbf{r} + \sum_{l \neq k} \frac{N_l^c}{R_{kl}} \qquad (2.32)$$

and eq. (2.31) becomes

$$E^{mol} = \frac{1}{\gamma^v} \sum_k \left[-Z_k^{eff} \int^{val} \frac{\rho(\mathbf{r})}{|\mathbf{r} - \mathbf{R}_k|}\, d\mathbf{r} + Z_k^{eff} \sum_{l \neq k} \frac{Z_l^{eff}}{R_{kl}} \right] + \sum_k E_k^{ion}. \qquad (2.33)$$

---

[1] Direct calculations [7] made for $1s$ electrons confirm the validity of eq. (2.29).

Here we observe that

$$\sum_k -Z_k^{\text{eff}} \int^{val} \frac{\rho(\mathbf{r})}{|\mathbf{r} - \mathbf{R}_k|} \, d\mathbf{r} = V_{\text{ne}}^{\text{eff}} \tag{2.34}$$

appropriately describes the total *effective* nuclear–electronic potential energy of the molecule, i.e., the potential energy of its valence electrons—and only those—in the field of the effective nuclear charges Z_k^{eff}, Z_l^{eff}, ... The summation of the nuclear repulsion terms gives $2V_{\text{nn}}^{\text{eff}}$, where $V_{\text{nn}}^{\text{eff}}$ is the total repulsion energy between Z_k^{eff}, Z_l^{eff}, ... etc., i.e.,

$$V_{\text{nn}}^{\text{eff}} = \sum_k \sum_{l>k} \frac{Z_k^{\text{eff}} Z_l^{\text{eff}}}{R_{kl}}. \tag{2.35}$$

The final result obtained from (2.33)–(2.35) is

$$E^{\text{mol}} = \frac{1}{\gamma^{\text{v}}}(V_{\text{ne}}^{\text{eff}} + 2V_{\text{nn}}^{\text{eff}}) + \sum_k E_k^{\text{ion}}. \tag{2.36}$$

This concludes the description of the core–valence separation in molecules. A sensible question is: what are we going to do with all this?

2.6   Perturbation of the valence region

Consider eq. (2.36). Consider also the valence-region energy of the isolated atoms, $E_k^{\text{v,at}}$, and rearrange (2.36) as follows:

$$\sum_k E_k^{\text{ion}} + \sum_k E_k^{\text{v,at}} - E^{\text{mol}} = \sum_k E_k^{\text{v,at}} - \frac{1}{\gamma^{\text{v}}}(V_{\text{ne}}^{\text{eff}} + 2V_{\text{nn}}^{\text{eff}}).$$

The total energy of an isolated ground-state atom k is $E_k^{\text{at}} = E_k^{\text{ion}} + E_k^{\text{v,at}}$ (Chapter 1.4). Equation (2.19) tells us that the left-hand side of this equation is ΔE_{a}^*, the energy of atomization. So we write

$$\Delta E_{\text{a}}^* = \sum_k E_k^{\text{v,at}} - \frac{1}{\gamma^{\text{v}}}(V_{\text{ne}}^{\text{eff}} + 2V_{\text{nn}}^{\text{eff}}). \tag{2.37}$$

ΔE_{a}^* is the central quantity involved in any comparison with thermochemical data, such as enthalpies of formation, enthalpies of atomization, and the like, and gives the total energy of a molecule with the help of (2.19). ΔE_{a}^* is what we want to calculate.

Suppose we are able to solve (2.37) in an approximate manner by making some assumptions—namely as regards model electron densities, $\rho^\circ(\mathbf{r})$, and internuclear distances, R_{kl}°, taken from a suitable reference system—and write

$$\Delta E_a^{*\circ} = \sum_k E_k^{v,at} - \frac{1}{\gamma^v}(V_{ne}^{eff} + 2V_{nn}^{eff})^\circ. \qquad (2.38)$$

The superscript "o" identifies values computed for the model. Combining (2.37) and (2.38) we have

$$\Delta E_a^* = \Delta E_a^{*\circ} - \frac{1}{\gamma^v}\Delta(V_{ne}^{eff} + 2V_{nn}^{eff}). \qquad (2.39)$$

It is usually not difficult to construct reasonable approximations for ΔE_a^*—this is what popular approximate bond-additivity schemes do[2]. So we have $\Delta E_a^{*\circ}$. Equation (2.39) indicates that the true ΔE_a^* can be obtained from the perturbation

$$\Delta(V_{ne}^{eff} + 2V_{nn}^{eff}) = (V_{ne}^{eff} + 2V_{nn}^{eff}) - (V_{ne}^{eff} + 2V_{nn}^{eff})^\circ \qquad (2.40)$$

describing the replacement of model densities, $\rho^\circ(\mathbf{r})$, and model internuclear distances, R_{kl}°, by the values which are appropriate for the molecule under scrutiny. Cleverly selected references require small corrections. Equations (2.39) and (2.40) and the idea of keeping the perturbation as small as possible disclose the strategy adopted in Chapter 3. Nature helps a lot in that matter by keeping the changes of $\rho(\mathbf{r})$ as small as possible.

The evaluation of $\Delta E_a^{*\circ}$ is made possible by the atom-by-atom partitioning of the molecular binding energy described by eq. (2.20). It was made clear at the outset, but is worth repeating now, that this sort of energy partitioning does not involve any spatial partitioning of a molecule into subspaces. (Subspaces defined by appropriate boundary surfaces can be of great interest, however. The strong persistence of recognizable atomic features in molecules is most vividly described in Bader's work [6] and is worth much more than passing attention.) It is true that any partitioning scheme necessarily requires a suitable criterion and raises questions about its uniqueness with respect to one or another fundamental postulate of quantum mechanics. The present approach towards core–valence separations does not severely confront this sort of problems if we accept the idea that the results obtained for the atoms are, at least to a good approximation, transferable to the molecules.

---

[2]We shall proceed with theoretical values, of course. *Noblesse oblige.*

2.7 Conclusion

The partitioning of the electronic charge of a molecule into core and valence parts unfolds as a straightforward extension of the methods applied to atoms.

It should be pointed out that, as it is the case with the isolated atoms, this partitioning expresses the valence-region energy of a molecule by reference to the collection of the *ground-state* ions (e.g., H^+, C^{+4}, N^{+5}, etc.) which would be left over upon removal of the valence-region electronic charge. This means that our formulas for the valence-region energy account, as they should, for core-relaxation effects.

Two results are especially important. On the one hand, there is equation (2.18) which describes the binding of an atom in a molecule: it will be used in the construction of appropriate reference systems permitting a simple evaluation of reference atomization energies, $\Delta E_a^{*\circ}$. On the other hand, we have equation (2.39) which permits to bridge the gap between the model atomization energy $\Delta E_a^{*\circ}$ and the true atomization energy ΔE_a^*. This is where we shall learn the most about the chemistry of chemical bonds. We take this equation with us when we begin Chapter 3.

Bibliography

[1] W. Kauzmann, "Quantum Chemistry", Academic Press, New York, 1957 (p. 229).

[2] R. E. Stanton, *J. Chem. Phys.*, **36**, 1298 (1962).

[3] R. G. Parr and S. R. Gadre, *J. Chem. Phys.*, **72**, 3639 (1980).

[4] S. Fliszár, M. Foucrault, M.-T. Béraldin, and J. Bridet, *Can. J. Chem.*, **59**, 1074 (1981).

[5] P. Politzer, *J. Chem. Phys.*, **64**, 4239 (1976); **69**, 491 (1978).

[6] R. F. W. Bader, "Atoms in Molecules, a Quantum Theory", Clarendon Press, Oxford, 1990.

[7] M. E. Schwartz, *Chem. Phys. Lett.*, **6**, 631 (1970).

Chapter 3

The Chemical Bond

3.1 Introduction

We have just found that the equation

$$\Delta E_a^* = \Delta E_a^{*o} - \frac{1}{\gamma^v}\Delta(V_{ne}^{eff} + 2V_{nn}^{eff}) \tag{3.1}$$

offers a sensible route towards the calculation of chemical binding. It is not the only way, of course, but it has the merit of offering a fresh insight into problems of chemical interest: bonds and bond energies. Bond dissociation energies are even more important, for that matter, because of their role in the interpretation of chemical reactions, but an understanding of bond energies in unperturbed ground-state molecules is required in the first place. Here we put emphasis on the factors governing the energies—and, subsequently, the dissociation—of chemical bonds without suggesting (even remotely) that eq. (3.1) is a panacea for solving all our problems. First of all, we should remember that (3.1) applies only to molecules at equilibrium.

There are other shortcomings with (3.1). This equation reveals only the electrostatic facets of our problem. The Hellmann–Feynman theorem tells us that all forces in a molecule can be understood on purely classical grounds, provided that the exact electron density (or at least a density derived from a wave function satisfying this theorem) is known for that molecule [1]. In this vein, the description of ΔE_a^* was reduced to purely electrostatic problems involving only nuclear–electronic and nuclear–nuclear interactions. But there is a price to be payed: the *mechanism* of bond formation cannot be understood in purely electrostatic terms [2, 3]. The kinetic energy of the electrons (which prevents them from collapsing into the nuclei) plays a decisive role because the electronic Hamiltonian of an atom or a molecule is only bounded from below if the kinetic energy is duly accounted for. The potential alone is not bounded from below. This decisive role deeply reflects in the theory explaining why chemical bonds are formed in the first place [2]. Electrostatic considerations are also important, of course, but do not suffice in that particular area. This important topic was masterfully covered by Kutzelnigg [3].

In short, eq. (3.1) allows no inquiry into the origin of chemical bonds. It is well suited, however, for describing properties pertaining to chemical bonds as they are found in molecules at equilibrium.

3.2   Bonded and nonbonded interactions

Under laboratory conditions, molecules contain vibrational, translational and rotational energies which are not fairly partitionable among chemical bonds. For a fair discussion of chemical binding it is appropriate to consider the molecules at their potential minimum, i.e., in their hypothetical vibrationless state at 0 K. The measure for what holds the atoms together in a molecule is thus given by ΔE_a^*, eq. (2.19), i.e., in conventional notation,

$$\Delta E_a^* = \sum_k \langle \Psi_k | \hat{H}^{at} | \Psi_k \rangle - \langle \Psi^{mol} | \hat{H}^{mol} | \Psi^{mol} \rangle. \tag{3.2}$$

It is understood that all the energies encountered in the following description refer exclusively to vibrationless ground-state molecules at 0 K.

At this stage we introduce the concept of *chemical bond* in the usual 'chemical' acceptation of this term. For any pair of bonded atoms k and l we speak in all cases of *one* bond only, whether this bond is a single one (in conventional language), like the CC bond in ethane, or a multiple bond (like the CC bonds in ethylene, acetylene, benzene, for example). This is a model, of course. This model translates familiar views: we regard a molecule as a collection of chemical bonds and associate an energy with each bond, e.g., ε_{kl} with the bond formed by atoms k and l. But it must be borne in mind that atom pairs which are not connected by a conventional chemical bond also interact somehow. The atomization energy thus includes the annihilation of all interactions between atom pairs which are not chemically bonded to one another in the conventional sense, ΔE_{nb}, and the destruction of all chemical bonds kl, represented by their energies ε_{kl}, i.e.,

$$\Delta E_a^* = \sum_{k<l} \varepsilon_{kl} + \Delta E_{nb}. \tag{3.3}$$

This model is closely tied to chemical ideas of structure. Each ε_{kl} represents the part of the total atomization energy that is contributed by a particular bond kl. This contribution is dictated by the actual properties of that bond in the molecule, namely, by the internuclear distance between atoms k and l and by their local charge densities. Each bond is part of the molecular fabric and its energy cannot be observed in isolation. This bond energy should not be mistaken for the energy required to break that bond. (The relationship between dissociation energies, D_{kl}, and intrinsic bond energies, ε_{kl}, is given in Chapter 4.) Here we confine ourselves to chemical bonds as they are found in the unperturbed ground-state molecules.

It is important to clarify what we mean by 'nonbonded interactions'. Consider the electronic charge density, $\rho(\mathbf{r})$, at any point $\mathbf{r}$ in the molecule. This density is a property of the entire molecule: in the neighborhood of any given atom, say, k, this density also depends on the presence of atoms not bonded to k. Broadly speaking, one could view this involvement of distant atoms as a nonbonded effect, but no such discrimination is made here. Anytime long-range (nonbonded) interactions interfere with the local densities of atoms k and l engaged in a bond we include this effect in the bond energy ε_{kl} and do not single out what is due to the presence of distant atoms. Briefly, distant atoms influence the energy of the bonds formed by atoms k and l by acting on their charge densities, but this is accounted for in the bond energy ε_{kl} and is not part of our definition of nonbonded interactions. Our definition of nonbonded interactions is the most common one and applies only to direct interactions between nonbonded atoms, say, r and s, resulting from all cross-interactions between their nuclei and their electrons—a definition which commands adequate knowledge about the electron populations associated with r and s.

Now consider the ΔE_{nb} term in eq. (3.3). It represents the difference between the nonbonded energy of the collection of isolated atoms (which is 0) and the nonbonded energy of the molecule, E_{nb}. This means that

$$\Delta E_{nb} = -E_{nb}. \tag{3.4}$$

Thus we rewrite (3.3) as follows

$$\Delta E_a^* = \sum_{k<l} \varepsilon_{kl}(\rho) - E_{nb}(\rho) \tag{3.5}$$

and specify that all quantities are calculated for the correct dentity ρ. Then we repeat the calculation using a model density ρ° and obtain the approximate result

$$\Delta E_{\mathrm{a}}^{*\circ} = \sum_{k<l} \varepsilon_{kl}(\rho^\circ) - E_{\mathrm{nb}}(\rho^\circ). \tag{3.6}$$

Now we are ready to work with eq. (3.1). Use (3.6) and write

$$\Delta E_{\mathrm{a}}^{*} = \sum_{k<l} \varepsilon_{kl}(\rho^\circ) - \frac{1}{\gamma^{\mathrm{v}}}\Delta(V_{\mathrm{ne}}^{\mathrm{eff}} + 2V_{\mathrm{nn}}^{\mathrm{eff}}) - E_{\mathrm{nb}}(\rho^\circ). \tag{3.7}$$

The $(1/\gamma^{\mathrm{v}})\Delta(V_{\mathrm{ne}}^{\mathrm{eff}} + 2V_{\mathrm{nn}}^{\mathrm{eff}})$ part of eq. (3.7) represents the correction that accompanies the change from ρ° to ρ and from R_{kl}° to R_{kl}. This correction concerns both the bonded and the nonbonded energy terms, a circumstance suggesting the following separation

$$\Delta(V_{\mathrm{ne}}^{\mathrm{eff}} + 2V_{\mathrm{nn}}^{\mathrm{eff}}) = \Delta(V_{\mathrm{ne}}^{\mathrm{eff}} + 2V_{\mathrm{nn}}^{\mathrm{eff}})^{\mathrm{bd}} + \Delta(V_{\mathrm{ne}}^{\mathrm{eff}} + 2V_{\mathrm{nn}}^{\mathrm{eff}})^{\mathrm{nb}} \tag{3.8}$$

where "bd" = bonded and "nb" = nonbonded. The first right-hand side term represents the correction that applies to a molecule viewed as a rigid assembly of chemical bonds and is amenable to a bond-by-bond decomposition (Section 4) because nonbonded interactions are entirely disregarded at this point. The last term of (3.8) corrects the nonbonded part of $\Delta E_{\mathrm{a}}^{*\circ}$. So we rewrite (3.7) as follows

$$\Delta E_{\mathrm{a}}^{*} = \sum_{k<l} \varepsilon_{kl}(\rho^\circ) - \frac{1}{\gamma^{\mathrm{v}}}\Delta(V_{\mathrm{ne}}^{\mathrm{eff}} + 2V_{\mathrm{nn}}^{\mathrm{eff}})^{\mathrm{bd}} - [\frac{1}{\gamma^{\mathrm{v}}}\Delta(V_{\mathrm{ne}}^{\mathrm{eff}} + 2V_{\mathrm{nn}}^{\mathrm{eff}})^{\mathrm{nb}} + E_{\mathrm{nb}}(\rho^\circ)]$$

and compare this expression with (3.5). We see that

$$\sum_{k<l} \varepsilon_{kl}(\rho) = \sum_{k<l} \varepsilon_{kl}(\rho^\circ) - \frac{1}{\gamma^{\mathrm{v}}}\Delta(V_{\mathrm{ne}}^{\mathrm{eff}} + 2V_{\mathrm{nn}}^{\mathrm{eff}})^{\mathrm{bd}} \tag{3.9}$$

$$E_{\mathrm{nb}}(\rho) = E_{\mathrm{nb}}(\rho^\circ) + \frac{1}{\gamma^{\mathrm{v}}}\Delta(V_{\mathrm{ne}}^{\mathrm{eff}} + 2V_{\mathrm{nn}}^{\mathrm{eff}})^{\mathrm{nb}}. \tag{3.10}$$

Equation (3.9) [4] is going to solve our problem of bond energies. Equation (3.10) solves nothing, unfortunately. It is in principle always possible to define a set of reference bond energies, $\{\varepsilon_{kl}(\rho^\circ)\}$, to construct a model molecule with these bonds, and to use eq. (3.9). But we cannot define a reference nonbonded energy because this 'reference' would change from molecule to molecule. In a way, E_{nb} is a stranger in the theory of chemical bonds.

In the spirit of our definition of nonbonded interactions, Del Re has shown that a valid approximation in σ systems is Coulombic in nature [5], i.e.,

$$E_{nb} = \frac{1}{2} \sum_{r,s}^{nb} \frac{q_r q_s}{R_{rs}} \tag{3.11}$$

where q_r and q_s are the *net* (nuclear minus electronic) charges of nonbonded atom pairs at a distance R_{rs}. Numerical evaluations [4, 6] indicate that $|E_{nb}| \ll \Delta E_a^*$, typically ~ 0.01–0.05% of ΔE_a^*. The chemical bonds, eq. (3.9), govern the major part, by far, of the properties determining ΔE_a^*.

3.3 Reference bonds

Equation (3.2) is our starting point. Using the Hellmann–Feynman theorem we get

$$\left(\frac{\partial \Delta E_a^*}{\partial Z_k} \right)_\rho = \left\langle \Psi_k \left| -\sum_i r_{ik}^{-1} \right| \Psi_k \right\rangle - \left\langle \Psi^{mol} \left| -\sum_i r_{ik}^{-1} + \sum_{l \neq k} \frac{Z_l}{R_{kl}} \right| \Psi^{mol} \right\rangle \tag{3.12}$$

where the first sum over i runs over all electrons of atom k with wave function Ψ_k. The second sum over i runs over all electrons of the molecule with wave function Ψ^{mol}. Finally, l runs over all nuclei but k and R_{kl} is the distance between nuclei k and l. Inspection of eq. (1.31) shows that the first integral is $V_{ne,k}/Z_k$—the potential at the nuclear position in the atom. The second integral is, from eq. (2.3), V_k/Z_k—the potential at the position of Z_k in the molecule. Equation (3.12) thus tells us that

$$V_k = V_{ne,k} - Z_k \left(\frac{\partial \Delta E_a^*}{\partial Z_k} \right)_\rho . \tag{3.13}$$

On the other hand, it follows from eq. (3.5) that

$$\left(\frac{\partial \Delta E_a^*}{\partial Z_k} \right)_\rho = \sum_l \left(\frac{\partial \varepsilon_{kl}}{\partial Z_k} \right)_\rho - \left(\frac{\partial E_{nb}}{\partial Z_k} \right)_\rho . \tag{3.14}$$

Combining (3.13) and (3.14) we get

$$V_k = V_{ne,k} - Z_k \sum_l \left(\frac{\partial \varepsilon_{kl}}{\partial Z_k} \right)_\rho + Z_k \left(\frac{\partial E_{nb}}{\partial Z_k} \right)_\rho . \tag{3.15}$$

Now use (1.36) with the appropriate labels $(k, l,$ etc.$)$, $E_k^{\text{at}} = (1/\gamma_k^{\text{at}})V_{\text{ne},k}$, as well as (2.15), $E_k^{\text{mol}} = (1/\gamma_k^{\text{mol}})V_k$, and calculate ΔE_k with the help of (2.18). This gives

$$\Delta E_k = \frac{Z_k}{\gamma_k^{\text{mol}}} \sum_l^{\text{bd}} \left(\frac{\partial \varepsilon_{kl}}{\partial Z_k}\right)_\rho + \frac{\gamma_k^{\text{mol}} - \gamma_k^{\text{at}}}{\gamma_k^{\text{mol}}} E_k^{\text{at}} - \frac{Z_k}{\gamma_k^{\text{mol}}} \left(\frac{\partial E_{\text{nb}}}{\partial Z_k}\right)_\rho. \qquad (3.16)$$

ΔE_k measures the binding of atom k in the molecule. It is obvious that the last term of the right-hand side of (3.16) has nothing to do with the chemical bonds formed by that atom. Only the first and second terms are involved in the decomposition of ΔE_k among the bonds formed by atom k.

The 'extraction' from the host molecule of an atom k forming v_k bonds requires an energy

$$\frac{1}{v_k} \cdot \frac{\gamma_k^{\text{mol}} - \gamma_k^{\text{at}}}{\gamma_k^{\text{mol}}} E_k^{\text{at}} \qquad (3.17)$$

for each bond, meaning that in the suppression of a kl bond this type of contribution must be counted once for atom k and once for atom l. In addition, the suppression of the kl bond requires an energy $(1/\gamma_k^{\text{mol}})Z_k(\partial \varepsilon_{kl}/\partial Z_k)_\rho$ and a similar energy for the partner l engaged in that bond. Consequently, the portion of the total atomization energy associated with the kl bond is, for $\varepsilon_{kl} = \varepsilon_{kl}(\rho)$,

$$\varepsilon_{kl} = \frac{Z_k}{\gamma_k^{\text{mol}}} \left(\frac{\partial \varepsilon_{kl}}{\partial Z_k}\right)_\rho + \frac{Z_l}{\gamma_l^{\text{mol}}} \left(\frac{\partial \varepsilon_{kl}}{\partial Z_l}\right)_\rho + \frac{1}{v_k} \cdot \frac{\gamma_k^{\text{mol}} - \gamma_k^{\text{at}}}{\gamma_k^{\text{mol}}} E_k^{\text{at}} + \frac{1}{v_l} \cdot \frac{\gamma_l^{\text{mol}} - \gamma_l^{\text{at}}}{\gamma_l^{\text{mol}}} E_l^{\text{at}}.$$
$$(3.18)$$

This is the sought-after expression for the energy of a chemical bond [4, 6]. Bond energies given by (3.18) satisfy exactly eq. (3.5).

The evaluation of the last two terms of the right-hand side of (3.18) causes no problems. The γ's and the atomic energies can be deduced from Hartree–Fock calculations [4, 6, 7]. One can also justify the use of experimental energies, a strategy which generates results rapidly converging towards experimental accuracy. The troublesome part is with the derivatives $(\partial \varepsilon_{kl}/\partial Z_k)_\rho$ and $(\partial \varepsilon_{kl}/\partial Z_l)_\rho$. The difficulties are revealed by inspection of (3.15). They can be overcome—with hard labor and some approximations— only in a few cases. The good news is that only a few ε_{kl}'s are needed, namely reference bond energies $\varepsilon_{kl}(\rho^\circ)$ calculated for model systems. Those determined for the CC and CH bonds of ethane, for example, are sufficient for the study of saturated hydrocarbons; the addition of the reference bond energy describing the double bond of ethene extends the range of applications to olefinic molecules, including polyenic material. It is thus well worth the trouble to calculate a few reference bond energies—and this can be done with reasonable accuracy—because the rest follows from applications of eq. (3.9). As a matter of fact, the use of (3.9) is far more instructive than numerical solutions of (3.18).

Equation (3.9) indicates that bond energies somehow depend on the distribution of electronic charge in the molecule because ΔV_{ne}^{eff} depends on ρ. Now, identical bond energies imply never changing internuclear distances between atoms that never change their electron populations. The point is that *invariant local electron populations cannot describe a set of electroneutral molecules*. Indeed, if constant electron populations N_k ($\neq Z_k$), N_l ($\neq Z_l$),...are associated with all individual atoms k, l,...of an electroneutral molecule, any other non-isomeric molecule constructed with the same atoms would not satisfy the charge-normalization requirement[1]. This argument suffices to put any additivity scheme postulating fixed bond energies on the disabled list. We must restore electroneutrality—a constraint introducing a charge dependence into the description of chemical bonds. In doing that we consider, as succinctly stated by Platt [8], that "a theory of chemistry and the chemical bond is primarily a theory of electron density".

3.4   The chemical bond

Now we go back to eq. (3.9). From here on we write ε_{kl}^o instead of $\varepsilon_{kl}(\rho^o)$ to designate reference bond energies and write ε_{kl} instead of $\varepsilon_{kl}(\rho)$. The electron densities, ρ, and the electron populations, N, refer to valence electrons. The $(1/\gamma^v)\Delta(V_{ne}^{eff} + 2V_{nn}^{eff})^{bd}$ term, of course, concerns only the interactions between bonded atoms. We shall momentarily assume—and this is quite an assumption—that we know how to assign electron populations N_k, N_l,...to individual 'atoms in a molecule', k, l, etc.

The contribution to ΔV_{ne}^{eff} involving Z_k^{eff} consists, first, of the interactions between Z_k^{eff} and the electrons with density ρ_k assigned to the atom k in the molecule, namely

$$- Z_k^{eff} \int^{\tau_k} \frac{\rho_k(\mathbf{r})}{|\mathbf{r} - \mathbf{R}_k|} \, d\mathbf{r} \qquad (3.19)$$

where the integration is carried out over the volume τ_k containing the N_k

---

[1]For example, if in C_2H_6 the C and H net charges are 0.0351 and -0.0117 e, respectively, a CH_4 molecule constructed with the same atoms would carry an excess negative charge of -0.0117 e.

electrons allocated to k, i.e.,

$$N_k = \int^{\tau_k} \rho_k(\mathbf{r}) \, d\mathbf{r}.$$

Second, Z_k^{eff} interacts with the N_l electrons of each atom l bonded to k, that is

$$- Z_k^{\text{eff}} \int^{\tau_l} \frac{\rho_l(\mathbf{r})}{|\mathbf{r} - \mathbf{R}_k|} \, d\mathbf{r} \tag{3.20}$$

with

$$N_l = \int^{\tau_l} \rho_l(\mathbf{r}) \, d\mathbf{r}.$$

The integrals in (3.19) and (3.20) are conveniently written

$$\int^{\tau_k} \frac{\rho_k(\mathbf{r})}{|\mathbf{r} - \mathbf{R}_k|} \, d\mathbf{r} = N_k \langle r_k^{-1} \rangle \tag{3.21}$$

and

$$\int^{\tau_l} \frac{\rho_l(\mathbf{r})}{|\mathbf{r} - \mathbf{R}_k|} \, d\mathbf{r} = N_l \langle r_{kl}^{-1} \rangle \tag{3.22}$$

where $\langle r_k^{-1} \rangle$ and $\langle r_{kl}^{-1} \rangle$ are the average inverse distances from Z_k^{eff} to N_k and N_l, respectively. The contribution involving Z_k^{eff} and its 'own' electrons N_k and the electrons N_l of every atom l forming a bond with k is, using (3.19)–(3.22),

$$V_{\text{ne},k}^{\text{eff}}(\rho) = - Z_k^{\text{eff}} [N_k \langle r_k^{-1} \rangle + \sum_l N_l \langle r_{kl}^{-1} \rangle]. \tag{3.23}$$

Here we specify that this expression holds for the true densities of the system under scrutiny. Similar expressions are written for the reference model with atomic populations N_k° and N_l° and average inverse distances $\langle r_k^{-1} \rangle^{\circ}$ and $\langle r_{kl}^{-1} \rangle^{\circ}$. This gives an equation like (3.23), but for model densities $\rho^{\circ}(\mathbf{r})$. Hence we obtain the difference

$$\Delta V_{\text{ne},k}^{\text{eff}} = V_{\text{ne},k}^{\text{eff}}(\rho) - V_{\text{ne},k}^{\text{eff}}(\rho^{\circ}). \tag{3.24}$$

Equation (3.24) is for nucleus Z_k^{eff}. Summation over all the nuclei and inclusion of the $2\Delta V_{\text{nn}}^{\text{eff,bd}}$ term gives

$$\Delta(V_{\text{ne}}^{\text{eff}} + 2V_{\text{nn}}^{\text{eff}})^{\text{bd}} =$$
$$- \sum_k Z_k^{\text{eff}} \left[N_k \langle r_k^{-1} \rangle - N_k^{\circ} \langle r_k^{-1} \rangle^{\circ} + \sum_l \left(N_l \langle r_{kl}^{-1} \rangle - N_l^{\circ} \langle r_{kl}^{-1} \rangle^{\circ} \right) \right]$$
$$+ \sum_k \sum_l Z_k^{\text{eff}} Z_l^{\text{eff}} \left[R_{kl}^{-1} - (R_{kl}^{-1})^{\circ} \right]. \tag{3.25}$$

What this equation says is simply that the addition of a small amount of electronic charge to an atom k modifies its 'own' atomic nuclear–electronic interaction energy, i.e., that involving Z_k, and, moreover, also changes the nuclear–electronic interactions with the nuclei of all atoms l bonded to k. The effects involving all the other atoms of the molecule are, of course, included in the nonbonded interaction term, E_{nb}. From here on, we transform (3.25) into something more practical, but equations become longer before they get shorter.

First we replace $\langle r_{kl}^{-1}\rangle$ by $\langle r_{kl}^{-1}\rangle^\circ$ where it occurs in (3.25) and restore the correct result in the following manner:

$$\Delta(V_{ne}^{eff} + 2V_{nn}^{eff})^{bd} = \tag{3.26}$$
$$-\sum_k Z_k^{eff}\left[N_k\langle r_k^{-1}\rangle - N_k^\circ\,\langle r_k^{-1}\rangle^\circ + \sum_l\left(N_l\langle r_{kl}^{-1}\rangle^\circ - N_l^\circ\langle r_{kl}^{-1}\rangle^\circ\right)\right]$$
$$+\sum_k\sum_l Z_k^{eff}Z_l^{eff}\left[R_{kl}^{-1} - (R_{kl}^{-1})^\circ\right] - \sum_k\sum_l Z_k^{eff}\left(N_l\langle r_{kl}^{-1}\rangle - N_l\langle r_{kl}^{-1}\rangle^\circ\right).$$

Now we define

$$N_l = N_l^\circ + \Delta N_l \tag{3.27}$$

for use in (3.26). Then we use the result in eq. (3.9) and write

$$\sum_{k<l}\varepsilon_{kl} = \sum_{k<l}\varepsilon_{kl}^\circ + \frac{1}{\gamma^v}\sum_k Z_k^{eff}\left[\left(N_k\langle r_k^{-1}\rangle - N_k^\circ\langle r_k^{-1}\rangle^\circ\right) + \sum_l \Delta N_l\langle r_{kl}^{-1}\rangle^\circ\right] + F. \tag{3.28}$$

The function F consists of the last two terms of (3.26) divided by γ^v, with a change in sign. It is convenient to rewrite F using the definition of net atomic charge, $Z_l^{eff} - N_l = q_l$. After some algebra one obtains [6]

$$F = -\frac{1}{\gamma^v}\sum_k\sum_l Z_k^{eff}Z_l^{eff}\left[R_{kl}^{-1} - \left(R_{kl}^{-1}\right)^\circ - \left(\langle r_{kl}^{-1}\rangle - \langle r_{kl}^{-1}\rangle^\circ\right)\right]$$
$$-\frac{1}{\gamma^v}\sum_k\sum_l Z_k^{eff}q_l\left(\langle r_{kl}^{-1}\rangle - \langle r_{kl}^{-1}\rangle^\circ\right). \tag{3.29}$$

F represents the contribution due to variations of internuclear distances and to changes of electronic centers of charge. The first part, in square brackets, is obviously 0 for spherically symmetric electron clouds and, more generally, if the centers of electronic charge move along with the nuclei during small changes in internuclear distances. The last part is small because atomic net charges are small in the first place (e.g., ≤ 0.0351 e for carbon) and because small changes of electron populations are unlikely to modify their center of charge to any significant extent. $F = 0$ proves accurate in applications to σ systems. Additional information concerning the use of (3.29) is given in Section 6.

Let us now consider the difference $N_k \langle r_k^{-1} \rangle - N_k^\circ \langle r_k^{-1} \rangle^\circ$ appearing in (3.28) and expand $N_k \langle r_k^{-1} \rangle$ in a Taylor series

$$N_k \langle r_k^{-1} \rangle = N_k^\circ \langle r_k^{-1} \rangle^\circ + \left(\frac{\partial N_k \langle r_k^{-1} \rangle}{\partial N_k} \right)^\circ \Delta N_k + \frac{1}{2!} \left(\frac{\partial^2 N_k \langle r_k^{-1} \rangle}{\partial N_k^2} \right)^\circ (\Delta N_k)^2 + \cdots$$

(3.30)

Then we define the energy

$$E_k^{vs} = -\frac{1}{\gamma_k^{mol}} Z_k^{eff} N_k \langle r_k^{-1} \rangle$$

(3.31)

of atom k in its valence state, in the current acceptation of this term, chosen so as to have the same interaction between the electrons of the atom as occurs when the atom is part of a molecule. The valence state is considered as being formed from a molecule by removing all the other atoms without allowing any electronic rearrangement in the atom of interest. It resembles the energy of (2.15), except that all interactions due to particles outside the volume τ_k are 'turned off' in E_k^{vs}. Taking now the successive derivatives of E_k^{vs} evaluated for $N_k = N_k^\circ$, i.e.,

$$\left(\frac{\partial E_k^{vs}}{\partial N_k} \right)^\circ = -\frac{Z_k^{eff}}{\gamma_k^{mol}} \left(\frac{\partial N_k \langle r_k^{-1} \rangle}{\partial N_k} \right)^\circ$$

$$\left(\frac{\partial^2 E_k^{vs}}{\partial N_k^2} \right)^\circ = -\frac{Z_k^{eff}}{\gamma_k^{mol}} \left(\frac{\partial^2 N_k \langle r_k^{-1} \rangle}{\partial N_k^2} \right)^\circ, \quad \text{etc.}$$

we obtain from (3.30) that

$$N_k \langle r_k^{-1} \rangle - N_k^\circ \langle r_k^{-1} \rangle^\circ = -\frac{\gamma_k^{mol}}{Z_k^{eff}} \left[\left(\frac{\partial E_k^{vs}}{\partial N_k} \right)^\circ \Delta N_k + \frac{1}{2!} \left(\frac{\partial^2 E_k^{vs}}{\partial N_k^2} \right)^\circ (\Delta N_k)^2 + \cdots \right]$$

(3.32)

Up to here, it was convenient to carry out the calculations using the valence-electron populations N_k, N_l, etc., as variables. It is more practical, however, to express the final results using *net* (i.e., nuclear minus electronic) charges, so that:

$$\Delta q = -\Delta N.$$

(3.33)

With this change, one obtains from (3.28), (3.32) and (3.33) that

$$\sum_{k<l} \varepsilon_{kl} = \sum_{k<l} \varepsilon_{kl}^{\circ}$$
$$+ \frac{1}{\gamma^{v}} \sum_{k} \left\{ \gamma_{k}^{\text{mol}} \left[\left(\frac{\partial E_{k}^{vs}}{\partial N_{k}} \right)^{\circ} \Delta q_{k} - \frac{1}{2!} \left(\frac{\partial^{2} E_{k}^{vs}}{\partial N_{k}^{2}} \right)^{\circ} (\Delta q_{k})^{2} + \cdots \right] \right.$$
$$\left. - Z_{k}^{\text{eff}} \sum_{l} \langle r_{kl}^{-1} \rangle^{\circ} \Delta q_{l} \right\} + F. \tag{3.34}$$

This equation contains the required information explaining the role of atomic charges in energy calculations. Molecular electroneutrality, i.e., $\sum_{k} q_{k} = 0$, is ensured with the use of the appropriate Δq's. Equation (3.34) is not practical in its present form. The following transformations are made with two goals in mind. First, we want to isolate those terms that need not be recalculated over and over again in each application of (3.34). Second, we want to cast (3.34) in a more instructive form, a form that highlights the properties of the individual bonds.

Final energy formulas

Let us define the quantity

$$a_{kl} = \frac{1}{v_{k}} \cdot \frac{\gamma_{k}^{\text{mol}}}{\gamma^{v}} \left[\left(\frac{\partial E_{k}^{vs}}{\partial N_{k}} \right)^{\circ} - \frac{1}{2!} \left(\frac{\partial^{2} E_{k}^{vs}}{\partial N_{k}^{2}} \right)^{\circ} \Delta q_{k} + \cdots \right] - \frac{1}{\gamma^{v}} Z_{l}^{\text{eff}} \langle r_{kl}^{-1} \rangle^{\circ} \tag{3.35}$$

where v_{k} = number of atoms attached to k. Multiply a_{kl} by Δq_{k} and carry out the double sum $\sum_{k} \sum_{l} a_{kl} \Delta q_{k}$. Comparison with (3.34) tells us that

$$\sum_{k<l} \varepsilon_{kl} = \sum_{k<l} \varepsilon_{kl}^{\circ} + \sum_{k} \sum_{l} a_{kl} \Delta q_{k} + F. \tag{3.36}$$

This form is most convenient in numerical calculations because the second-order term $\frac{1}{2}(\partial^{2} E_{k}^{vs}/\partial N_{k}^{2})^{\circ} \Delta q_{k}$ (and all the higher-order terms) can be neglected in most cases and the a_{kl}'s can be treated and tabulated like constants. The higher-order terms can always be reintroduced when needed.

The second transformation concerns $\sum_{k<l} \varepsilon_{kl}^{\circ} + F$ and is best explained by an example. Suppose we are studying saturated hydrocarbons. This requires two reference bond energies, namely, ε_{CC}° for the bonds between sp^{3} carbons and ε_{CH}° for the CH bonds. These two reference energies describe the original set, $\{\varepsilon_{kl}^{\circ}\}_{\text{original}}$, for use in eq. (3.36) and can be chosen to make

$F = 0$ for this class of compounds. Now we include a reference energy describing a carbon–carbon double bond in that original set and use it in applications to unsaturated hydrocarbons like ethene, propene, etc. We find that F is not any longer 0. This is due to the presence of new types of bonds, not found in $\{\varepsilon_{kl}^{\circ}\}_{\text{original}}$, namely, the $C(sp^2)$–$C(sp^3)$ and $C(sp^2)$–H bonds. One possible strategy is to keep using $\{\varepsilon_{kl}^{\circ}\}_{\text{original}}$ and to calculate F by means of (3.29). The other strategy involves a decomposition of F that permits a definition of new references, specifically tailored for $C(sp^2)$–$C(sp^3)$ and $C(sp^2)$–H bonds. (Section 6 describes how this is done.) In this manner on obtains a 'modified set', so that

$$\left\{ \sum_{k<l} \varepsilon_{kl}^{\circ} \right\}_{\text{original}} + F = \left\{ \sum_{k<l} \varepsilon_{kl}^{\circ} \right\}_{\text{modified}} . \tag{3.37}$$

Clearly, the reference bond for $C(sp^2)$–$C(sp^3)$ is derived from that representing $C(sp^3)$–$C(sp^3)$ and $C(sp^2)$–H is derived from $C(sp^3)$–H by incorporating the appropriate parts of F into the new reference energies. This revised definition of reference bonds does not alter the meaning of (3.36), but only its form: it is justified because chemists usually discriminate between different types of bonds. If we stipulate that $\sum_{k<l} \varepsilon_{kl}^{\circ}$ represents such a revised set, we can rewrite (3.36) as follows:

$$\sum_{k<l} \varepsilon_{kl} = \sum_{k<l} \varepsilon_{kl}^{\circ} + \sum_{k} \sum_{l} a_{kl} \Delta q_k . \tag{3.38}$$

It is now easy to separate (3.38) into individual bond contributions. This gives the final energy formula for a chemical bond [4, 6], that is

$$\varepsilon_{kl} = \varepsilon_{kl}^{\circ} + a_{kl} \Delta q_k + a_{lk} \Delta q_l . \tag{3.39}$$

This equation vividly shows the role played by the charges of the bond-forming atoms k and l: ε_{kl}° is for a reference bond with net charges q_k° and q_l° at atoms k and l, respectively, whereas ε_{kl} corresponds to modified charges, $q_k = q_k^{\circ} + \Delta q_k$ and $q_l = q_l^{\circ} + \Delta q_l$. The parameters a_{kl} and a_{lk}, 'measuring' the changes in bond energy accompanying unit charge variations at atoms k and l, respectively, are deduced from eq. (3.35). The total atomization energy is obtained from (3.5) and (3.39). Equation (3.39) is at the heart of the theory of bond dissociation energies.

A final comment is in order. In this Section we considered for the first time a partitioning describing a molecule as a collection of subspaces with volumes τ_k, τ_l,...containing N_k, N_l,...electrons, respectively. Briefly, we introduced the idea that electron populations N_k, N_l, etc., can be assigned to individual atoms in a molecule. These populations are volume-integrated stationary electron densities and not necessarily integer numbers. No definition was given for the boundaries separating the subspaces. This could be done, e.g., by accepting Bader's criteria [9] of space partitioning, but is not considered here. Thus, no idea regarding specific properties of atomic subsystems has developed (or has been invoked), say, whether or not the virial theorem is locally satisfied in each subspace. In the present approach, the problem of space partitioning has been reduced to a calculation of numbers, the variations of atomic charges—a problem that can be treated by Mulliken-type population analyses (Chapter 5)[2].

3.5 Theoretical vs. empirical bond energies

To help develop a familiarity with eq. (3.39), we shall derive and discuss a general formula for saturated hydrocarbons, $C_nH_{2n+2-2m}$, containing m six-membered rings ($m = 0, 1, 2, \ldots$). These molecules contain $(n - 1 + m)$ CC bonds and $(2n + 2 - 2m)$ CH bonds; hence

$$\sum_{k<l} \varepsilon_{kl}^{\circ} = (n - 1 + m)\,\varepsilon_{CC}^{\circ} + (2n + 2 - 2m)\,\varepsilon_{CH}^{\circ}. \tag{3.40}$$

The reference bonds are those of ethane[3], namely, $\varepsilon_{CC}^{\circ} = 69.63$ and $\varepsilon_{CH}^{\circ} = 106.81$ kcal mol^{-1} for $q_C^{\circ} = 35.1$ and $q_H^{\circ} = -11.7$ me (1 me $= 10^{-3}$ electron).

Consider first the carbon atoms. Each carbon forms N_{CC} bonds with other carbon atoms and $4 - N_{CC}$ bonds with hydrogen atoms. The $\sum_l a_{kl}\Delta q_k$ part for that C atom is $N_{CC}a_{CC}\Delta q_C + (4 - N_{CC})a_{CH}\Delta q_C$. Each H atom contributes $a_{HC}\Delta q_H$. Summation over all C and H atoms gives

$$\sum_k \sum_l a_{kl}\Delta q_k = (a_{CC} - a_{CH})\sum N_{CC}\Delta q_C + 4a_{CH}\sum \Delta q_C + a_{HC}\sum \Delta q_H. \tag{3.41}$$

Charge normalization tells us i) that $\sum q_H = -\sum q_C$ and ii) that $q_C^{\circ} + 3q_H^{\circ} = 0$ (for ethane). Using these constraints in the calculation of $\sum \Delta q_H = \sum(q_H - q_H^{\circ})$ one finds

$$\sum \Delta q_H = -\sum \Delta q_C + (n - 2 + 2m)\,q_H^{\circ}$$

---

[2]This does not mean to imply that alternatives should not be explored.
[3]The numerical values are described in Chapter 6.

and (3.41) becomes

$$\sum_k \sum_l a_{kl}\Delta q_k = A_1 \sum N_{CC}\Delta q_C + A_2 \sum \Delta q_C + (n-2+2m)\,a_{HC}q_H^\circ \quad (3.42)$$

where $A_1 = a_{CC} - a_{CH}$ and $A_2 = 4a_{CH} - a_{HC}$. The sum (3.40)+((3.42) gives $\sum_{k<l}\varepsilon_{kl}$, the general solution for the $C_nH_{2n+2-2m}$ hydrocarbons. The appropriate parameters are: $a_{CC} = -0.488$, $a_{CH} = -0.247$ and $a_{HC} = -0.632$ kcal mol^{-1} me^{-1}.

Now we come to the point that deserves special attention. Remembering that $q_C^\circ + 3q_H^\circ = 0$, we can write the following identity:

$$(n - 2 + 2m)\,q_H^\circ = -\frac{1}{2}(n - 1 + m)\,q_C^\circ - \frac{1}{4}(2n + 2 - 2m)\,q_H^\circ.$$

Using this identity and combining (3.40) and (3.42) it appears that

$$\begin{aligned}
\sum_{k<l}\varepsilon_{kl} &= (n - 1 + m)(\varepsilon_{CC}^\circ - \frac{1}{2}a_{HC}q_C^\circ) + (2n + 2 - 2m)(\varepsilon_{CH}^\circ - \frac{1}{4}a_{HC}q_H^\circ) \\
&\quad + A_1 \sum N_{CC}\Delta q_C + A_2 \sum \Delta q_C.
\end{aligned} \quad (3.43)$$

Imagine someone makes a multiple regression using experimental atomization energies, $\Delta E_a^* \simeq \sum_{k<l}\varepsilon_{kl}$, and justifies departures from exact bond additivity by invoking structural effects (like butane-*gauche*-, skew-pentane, etc., interactions) as a substitute for $A_1 \sum N_{CC}\Delta q_C + A_2 \sum \Delta q_C$. The regression analysis would provide the coefficients of $(n - 1 + m)$ and $(2n + 2 - 2m)$—which represent the numbers of CC and CH bonds, respectively—and these coefficients would then be christened CC and CH bond energies. This is how empirical correlations work and how empirical bond energies are determined. If 'good' substitutes are used for $A_1 \sum N_{CC}\Delta q_C + A_2 \sum \Delta q_C$, it appears that

$$\varepsilon_{CC}(\text{empirical}) \simeq \varepsilon_{CC}^\circ - \frac{1}{2}a_{HC}q_C^\circ = 80.7 \quad \text{kcal mol}^{-1}$$

$$\varepsilon_{CH}(\text{empirical}) \simeq \varepsilon_{CH}^\circ - \frac{1}{4}a_{HC}q_H^\circ = 105 \quad \text{kcal mol}^{-1}$$

These are, indeed, the values to which we have been accustomed, with variants depending on the hypotheses involved in their derivation. Empirical bond energies of this sort have become an almost unerasable part of our grammar. It should be clear by now that they do not represent true bond energies. Theory explains their origin but does not justify their use.

3.6 Valence atomic orbital centroids

The analysis presented in Section 4 has primarily dealt with the *numbers* of electrons associated with the individual atoms in a molecule. Now we examine the *shape* of these electron populations. They are important in energy calculations. Of course, as in Section 4, electron densities, ρ, and electron populations, N, are those of the valence region.

Here we focus attention on the external electrostatic potential at atom k, i.e.,

$$\mathcal{V}_k = -\sum_l \left[\int \frac{\rho_l(\mathbf{r})}{|\mathbf{r} - \mathbf{R}_k|} \, d\mathbf{r} - Z_l^{\text{eff}} R_{kl}^{-1} \right]$$

where $\rho_l(\mathbf{r})$ is the density of the N_l valence electrons assigned to atom $l \neq k$, and Z_l^{eff} is the effective nuclear charge of that atom, at a distance R_{kl} from atom k. Using (3.22), we write

$$\mathcal{V}_k = -\sum_l \left(N_l \langle r_{kl}^{-1} \rangle - Z_l^{\text{eff}} R_{kl}^{-1} \right). \tag{3.44}$$

Let us also introduce the corresponding expression obtained when the electron populations are kept at their proper values in the molecule under study, but the inverse-distance terms are replaced by their values (indicated by the superscript "o") of a model reference molecule, i.e.,

$$\mathcal{V}_k^{\text{nr}} = -\sum_l \left[N_l \langle r_{kl}^{-1} \rangle^{\circ} - Z_l^{\text{eff}} \left(R_{kl}^{-1} \right)^{\circ} \right]. \tag{3.45}$$

The difference $\mathcal{V}_k - \mathcal{V}_k^{\text{nr}}$ describes a change in electrostatic potential at atom k. The corresponding change in potential energy, summed over all atoms, is

$$f = \sum_k Z_k^{\text{eff}} \left(\mathcal{V}_k - \mathcal{V}_k^{\text{nr}} \right). \tag{3.46}$$

This is the quantity we shall use for our discussion. It is a convenient way of gaining insight into the function F, eq. (3.29), because

$$f = -\gamma^{\text{v}} F. \tag{3.47}$$

Extensive numerical calculations (Chapter 6) indicate i) that f is negligible (say, < 0.3 kcal mol^{-1}) for saturated hydrocarbons but ii) significant for olefinic molecules (e.g., ~ 40 kcal mol^{-1} for tetramethylethylene). The condition that f should vanish can be satisfied either because the various atomic contributions to f cancel, or because the individual terms in the summation over k vanish. Since it seems unlikely that cancellation of terms associated to different atoms would take place systematically in a large number of molecules, we shall assume that, to a good approximation,

$$\mathcal{V}_k \simeq \mathcal{V}_k^{nr} \qquad (3.48)$$

for any atom of any saturated hydrocarbon.

The most direct interpretation of (3.48) follows from the observation, suggested by eqs. (3.44) and (3.45), that f is essentially a 'relaxation' term. In fact, $\mathcal{V}_k - \mathcal{V}_k^{nr}$ represents the difference between the electrostatic potential at the kth nucleus in the given molecule and the potential which the same nucleus would feel if the atomic orbitals and the equilibrium distances remained the same as in the reference molecule in spite of the change in electron populations. With this picture in mind, eq. (3.48) reads: "Whenever the atoms under consideration in a given molecule are in the same valence states as in the reference molecule, the relaxation process is such that the potential created by the other atoms at the kth nucleus is the same as would be predicted by leaving the pertinent internuclear distances and the shapes of atomic electron densities as they are in the reference molecule, the electron populations being changed as required by the new situation". This may mean that changes in the nuclear positions and in the centroids of the atomic orbitals always take place so as to leave the ratio between the expectation value $\langle r_{kl}^{-1} \rangle$ and R_{kl}^{-1} the same as in the reference molecule or, at least, that changes are insignificant. It is impossible to decide which alternative applies from a study of the paraffins, since their bond distances and atomic orbitals are expected to be practically constant.

A simple way of satisfying eq. (3.48) consists in assuming that

$$\langle r_{kl}^{-1} \rangle \simeq R_{kl}^{-1} \qquad (3.49)$$

when all bonds are σ bonds at their equilibium geometries, as in the alkanes. This simplifying hypothesis is supported by examination of SCF potentials at the nuclei, showing that (3.49) holds at least to within $\sim 5\%$, both for CH and CC bonds. It is especially useful here because it permits us to isolate the relevant conceptual points in a straightforward manner within the simple framework of the point-charge approximation. Defining a density $\rho_{\mu l}$ and an electron population $N_{\mu l}$ associated with each atomic orbital μ of atom l, such that

$$\int \frac{\rho_{\mu l}(\mathbf{r})}{|\mathbf{r} - \mathbf{R}_k|} \, d\mathbf{r} = N_{\mu l} \langle r_{\mu kl}^{-1} \rangle \qquad (3.50)$$

we can write[4]

$$\sum_\mu \frac{N_{\mu l}}{N_l} \langle r_{\mu k l}^{-1} \rangle = 1/R_{kl}. \tag{3.51}$$

Equation (3.51) reads: "The average of the expectation values of $|\mathbf{r} - \mathbf{R}_k|^{-1}$ for the various valence AO's of atom l, weighted by the ratios of the orbital populations to the total atomic population of atom l equals the inverse of the k–l distance". For all their simplicity, eqs. (3.50) and (3.51) cannot be tested numerically by direct calculation. The reason is linked to the difficulty of partitioning the total electron density into atomic contributions, in spite of an important conceptual step forward due to Parr [13]. A practical step in the same direction is the construction of suitable *in situ* valence atomic orbitals (VAO) from accurate ab initio computations [14], as advocated a long time ago by Mulliken [15] and discussed by Del Re [16]. As will be seen, such *in situ* VAO's do provide useful information, but they are of no help in solving the additional problem of assigning suitable populations to the orbitals and of dividing overlap populations into atomic contributions. In view of this situation, we take (3.48) and (3.51) as statements whose validity rests on experimental evidence, at least for saturated hydrocarbons.

In paraffins, the bonds formed by carbon are either CH or CC single bonds; of course, the electron density around C is not exactly spherical, but special multipole contributions appear to vanish to the extent that one can regard that the CH electron density adjusts to that of CC, and conversely, so that electrostatic effects are adequately described by a point charge located at C. Now suppose that one bond to hydrogen is removed and replaced by a π bond of the same atom to another carbon atom. Changes in geometry and hybridization will, of course, occur but, at least to first order, the resulting π atomic orbital will not be able to participate in a possible mutual readjustment of the AO centroids, due to its different symmetry. Thus we have to expect that condition (3.51) also applies to olefins, with the restriction that it only concerns σ orbitals. In other words, ethylenic carbon atoms will obey a 'principle' of σ-bond electrostatic balancing [17], that is

$$\begin{aligned} N_l \langle r_{kl}^{-1} \rangle &= \int \frac{\rho_l(\mathbf{r})}{|\mathbf{r} - \mathbf{R}_k|} \, d\mathbf{r} = \sum_\mu \int \frac{\rho_{\mu l}(\mathbf{r})}{|\mathbf{r} - \mathbf{R}_k|} \, d\mathbf{r} \\ &= \frac{N_l - N_{\pi l}}{R_{kl}} + N_{\pi l} \langle r_{\pi k l}^{-1} \rangle. \end{aligned} \tag{3.52}$$

---

[4]Equation (3.50) implies that orbital products can be well approximated in the average by some Mulliken-type expansion [5, 11]. This does not imply that the populations introduced are gross Mulliken populations [12].

Let us now consider $\langle r_{\pi kl}^{-1} \rangle$. A π atomic orbital is essentially a pure p orbital: if there is any polarization (as will be discussed below), this will involve a very small displacement, $\Delta r_{\pi l}$, of the centroid of the orbital. Then a monopole approximation of the π term of eq. (3.52) suffices, giving

$$\langle r_{kl}^{-1} \rangle = \frac{1 - N_{\pi l}/N_l}{R_{kl}} + \frac{N_{\pi l}/N_l}{|\mathbf{r}_l + \Delta \mathbf{r}_{\pi l} - \mathbf{R}_k|}. \tag{3.53}$$

If this simple expression is sufficient to account for the discrepancies observed in calculations of olefins based on alkane reference bonds—i.e., calculations using (3.36) with $F = 0$—we can claim that the 'principle' (3.51) is, indeed, satisfied by ethylenic hydrocarbons.

Turning now to direct theoretical evaluations, we consider $|\Delta \mathbf{r}_{\pi l}|$ as the displacement (on the C=C axis) of the centroid of the π orbital with respect to the center l. Of course, such a displacement can differ from zero only if some hybridization is allowed which, in the case of a π orbital, must consist in admixture of the suitable $d\pi$ orbital. The hybrid in question was determined [17] from 4–31G calculations with d polarization functions for carbon and optimization of all scale factors, followed by a calculation of *in situ* valence orbitals, and of their characteristics, according to Del Re and Barbier [14]. The inward shifts (on the C=C axis) of the π orbital centroids are close to 0.03 Å (Table 3.1).

Applications of eq. (3.53) are straightforward, using

$$|\mathbf{r}_l + \Delta \mathbf{r}_{\pi l} - \mathbf{R}_k| = \sqrt{R^2 + (\Delta r_{\pi l})^2 - 2R\,\Delta r_{\pi l} \cos\varphi} \tag{3.54}$$

where R is the length of the $C(sp^2)$–k bond (k = H or $C(sp^3)$) and φ is the angle it forms with the axis of the carbon-carbon double bond. Experimental geometries and the ethylene $C(sp^2)$ reference populations $N_{\pi l} = 1$ and $N_l = 3.9923$ e (Chapter 5) were used [17] to solve (3.53) with the help of (3.54).

TABLE 3.1. Inward shifts, $\Delta r_{\pi l}$ on the C=C
axis, of π orbital centroids

Molecule	Bond k–l	Δr (Å)
Ethylene	H–$C(sp^2)$	0.0292
Propene	H–$C_1(sp^2)$	0.0281
	H–$C_2(sp^2)$	0.0299
Tetramethylethylene	$C(sp^3)$–$C(sp^2)$	0.0280

TABLE 3.2. Atomization energies of olefins
calculated with and without inclusion of F

Molecule	ΔE_a^*, kcal mol^{-1}		
	$F = 0$	F_{included}	Exptl.
Ethylene	569.71	562.03	562.10
Propene	868.60	859.00	858.56
Tetramethylethylene	1763.89	1747.53	1746.64

This allows the calculation of F, eq. (3.29), where $\langle r_{kl}^{-1}\rangle^\circ = (R_{kl}^{-1})^\circ$, eq. (3.39), taking ethane as reference. (The approximation $\gamma^v = \frac{7}{3}$ was used.) Table 3.2 reports $\Delta E_a^* \simeq \sum_{k<l} \varepsilon_{kl}$ energies deduced from (3.36) (and the ethane reference bonds) with and without inclusion of the appropriate F values[5].

For all their simplicity, the ideas embodied in eq. (3.53) offer a realistic interpretation explaining the essence of the important modifications taking place in CC and CH bonds due to the replacement of an sp^3 hybridized carbon by an sp^2 carbon atom. These ideas materialize in the construction of individual bond contributions, F_{kl}, for each type of bond, that is, from eq. (3.29),

$$
F_{kl} = -\frac{1}{\gamma^v} Z_k^{\text{eff}} Z_l^{\text{eff}} \left[R_{kl}^{-1} - \left(R_{kl}^{-1}\right)^\circ - \left(\langle r_{kl}^{-1}\rangle - \langle r_{kl}^{-1}\rangle^\circ\right)\right]
$$
$$
- \frac{1}{\gamma^v} Z_k^{\text{eff}} q_l \left(\langle r_{kl}^{-1}\rangle - \langle r_{kl}^{-1}\rangle^\circ\right). \tag{3.55}
$$

Despite the marked differences both in geometrical parameters and in the SCF Δr values between the molecules involved in this comparison, there are striking regularities: the F value calculated for propene is, for all practical purposes, $\frac{3}{4}$ that of ethylene (which takes care of 3 CH bonds) plus $\frac{1}{4}$ that of tetramethylethylene (for the CC bond). Capitalizing on this idea, we can consider transferable bond contributions modelled after (3.55) and use them to generate new reference bond energies satisfying (3.37). This considerably simplified approach works extremely well in molecular calculations (Chapter 6). The bottom line is that we can safely proceed with eq. (3.39), so long as we use the appropriate reference bond energies ε_{kl}°, because these reference energies enjoy a high degree of transferability.

---

[5] Detailed examples illustrating the calculation of F are worked out in Chapter 6.3.

3.7   The calculation of pi systems

Consider the definition of a_{kl}, eq. (3.35). The numerical evaluation of a_{kl} involves the derivatives $(\partial E_k^{\mathrm{vs}}/\partial N_k)^\circ$, $(\partial^2 E_k^{\mathrm{vs}}/\partial N_k^2)^\circ$, etc. When atom k is an sp^2 carbon, we can safely neglect the second- and higher-order terms because the Δq_k's are small. However, we must consider both σ and π electron densities and their variations. The first derivatives $(\partial E_k^{\mathrm{vs}}/\partial N_k)^\circ$ calculated for ethylene are [18] -0.375 and -0.246 au for σ and π electrons, respectively. Moreover, considering the shift of the π orbital centroid, $\langle r_{kl}^{-1}\rangle^\circ$ also depends on whether we are describing σ or π electrons. Briefly, two a_{kl} parameters occur in unsaturated systems, namely $a_{kl\sigma}$ and $a_{kl\pi}$ for σ and π electrons, respectively.

Separating now the variations of the σ charges, $\Delta q_{k\sigma}$, from those of the π charges, $\Delta q_{k\pi}$, we write

$$a_{kl}\Delta q_k = a_{kl\sigma}\Delta q_{k\sigma} + a_{kl\pi}\Delta q_{k\pi}. \tag{3.56}$$

On the other hand, we have $\Delta q_k = \Delta q_{k\sigma} + \Delta q_{k\pi}$, meaning that the a_{kl} of eq. (3.56) is simply the weighted average of $a_{kl\sigma}$ and $a_{kl\pi}$, that is

$$a_{kl} = \frac{a_{kl\sigma}\Delta q_{k\sigma} + a_{kl\pi}\Delta q_{k\pi}}{\Delta q_{k\sigma} + \Delta q_{k\pi}}. \tag{3.57}$$

It turns out that the variations of σ and π charges are not independent variables [37]. They obey, at least to a good approximation, a simple relationship

$$\Delta q_{k\sigma} = m\Delta q_{k\pi} \tag{3.58}$$

where m (< 0) describes how σ populations at atom k increase when π populations decrease, and viceversa. Combining (3.57) and (3.58) we get

$$a_{kl} = \frac{ma_{kl\sigma} + a_{kl\pi}}{m + 1}. \tag{3.59}$$

The technical difficulty encountered in applications of eq. (3.59) comes from the fact that calculated σ charges appear to be considerably more basis set dependent than π charges, rendering the evaluation of m uncertain. Presently, the best estimates are $m \simeq -0.955$ for ethylenic molecules [6] and $m \simeq -0.814$ for benzenoid hydrocarbons [18], but these estimates should be taken with a grain of salt. It would be welcome if someone could figure out how to get better results.

3.8   Conjugation

The modifications taking place in CC and CH bonds due to the replacement of an sp^3 hybridized carbon by an sp^2 carbon atom and the explicit introduction of the σ-π separation are important in applications of the energy formula (3.36) to cover olefins. At this point we have considered the merits of the function F and dealt with mono-olefins applying the same scheme as used for paraffins. In applications to polyunsaturated systems, such as 1,3-butadiene and aromatic hydrocarbons, it appears that the function F does not translate all possible geometry-related effects. Attention must be given to an important property of π-systems—conjugation—which is introduced as follows, taking butadiene as an example.

The electron diffraction investigation of the molecular structure of 1,3-butadiene vapor indicates that the planar *trans* form is predominating: no signs of other conformations were observed [20], a result which is convincingly supported by ab initio MO theory [21], at the level of minimal (STO-3G) and extended (4-31G, 6-31G, 6-31G*, $(7s3p \mid 3s)$ and larger) basis set calculations. Structural parameters were determined [20, 22], namely $R(\text{C–C}) = 1.463 \pm 0.003$, $R(\text{C=C}) = 1.341 \pm 0.002$, and $R(\text{C–H}) = 1.090 \pm 0.004$ Å. While the double bond is, in essence, that of ethylene itself (a fact which prompts us to treat all double bonds on the same footing), the single CC bond of 1,3-butadiene is significantly shorter than the 'usual' single bonds, i.e., those formed by two sp^3 carbon atoms. The short CC single bond and the planar equilibrium conformation have been attributed to π-electron delocalization which gives the single bond some amount of double bond character and is most effective in the planar molecule [23]. It became apparent, however, that delocalization was less than predicted by the simple Hückel method [24] and the CC bond shortening was then attributed to the change in hybridization of the bond-forming atoms [25]. These points merit consideration in the interpretation of anticipated energy effects resulting from steric constraints that would force non-coplanarity of the double bond systems.

Extended basis set MO calculations indicate that, indeed, resonance is the main factor responsible for the shortening of the CC single bond in the planar *s-trans* conformation of 1,3-butadiene [26, 27]. In this conformation the distinction between σ and π electrons is evident. For the perpendicular case, on the other hand, it is always possible to define one localized π MO on each C=C fragment. The SCF localized π MO located on one of the C=C moieties has σ tails on the other C=C fragment and the σ MO's of each C=C moiety have components on the $2p_y$ atomic orbitals associated with the other fragment. This effect which is known as hyperconjugation has been found to stabilize the perpendicular form, largely compensating the lack of π conjugation, and leads to a similar shortening of the central CC bond [27].

Ab initio analyses thus substantiate the concept of π conjugation in the planar conformation and extend it to hyperconjugation in perpendicular forms. Direct calculations of resonance energies lead to estimates in the neighborhood of ~ 10 kcal mol^{-1} [21, 27]. Here, resonance energy is defined in its original sense as the energy difference between the conjugated system and its reference state without resonance, the latter being represented by a model wave function consisting of strictly localized nonresonating π molecular orbitals. Hyperconjugation energy is defined in similar manner [27]. The hyperconjugative stabilization in the perpendicular form at its SCF optimum distance (~ 1.5 Å) turns out to be almost as large (~ 8.0 kcal mol^{-1}) as the π conjugation energy in the equilibrium planar conformation (~ 10.4 kcal mol^{-1} at ~ 1.48 Å) [27]. This change from resonance to hyperconjugative stabilization of the central CC bond, however, is not reflected in the function F, nor is it in the $\sum_k \sum_l a_{kl}\Delta q_k$ term because the latter considers only the effects of changing electron populations at the bond-forming atoms under otherwise identical conditions. Rather, the change in geometry from a nonplanar to a planar form which is at the origin of the gain in conjugation at the expense of a hyperconjugative stabilization should be accounted for and taken care of by an appropriate change of the reference bond energy, ε_{CC}°, of the central CC single bond.

The extent of the change in ε_{CC}°, however, is difficult to evaluate theoretically. A valuable rough estimate is offered by the SCF results discussed above [27] indicating that hyperconjugative stabilization is less effective (by ~ 2.4 kcal mol^{-1}) than π conjugation. This estimate is similar to the familiar thermochemical resonance energy of some 3 kcal mol^{-1} [28] and suggests that the reference bond energy ε_{CC}° in a planar resonating arrangement should be larger by approximately that amount relative to that applicable in a markedly nonplanar situation. In practical applications, a uniform value of ~ 2.8 kcal mol^{-1} seems a reasonable, but perforce approximate, compromise.

While π conjugation or hyperconjugation are certainly important contributions to the thermochemical stability of CC single bonds embedded in double bond systems, molecules featuring this arrangement still appear to be strongly localized (i.e., poorly conjugated). Of course, this fact is well known from simple π-Hückel theory with equal CC resonance integrals leading to alternating π-bond orders. In this sense, keeping in mind the features linked to π conjugation *vs.* hyperconjugation revealed by SCF theory, it remains valid to apply our energy formula (3.36) to conjugated systems with the understanding that resonance effects must be taken into consideration.

3.9 Summary

The Hellmann–Feynman theorem offers a convenient way to bring out the main features of chemical binding: by taking the nuclear charges as parameters, it becomes possible to define the binding of each individual atom in a molecule without having recourse to a spatial partitioning of that molecule into atomic subspaces.

Consideration of the valence region of a molecule, modelled on that of the isolated atoms, leads to an expression describing the exact atomization energy in terms of an approximate solution based on a model and a perturbation with respect to the parameters postulated for that model. The final result is:

$$\Delta E_{\mathrm{a}}^* = \sum_{k<l} \varepsilon_{kl}^{\circ} + \sum_k \sum_l a_{kl}\Delta q_k + F - E_{\mathrm{nb}}. \qquad (3.60)$$

A simple sum of fixed reference bond energies, $\sum_{k<l} \varepsilon_{kl}^{\circ}$, does not, in general, describe an electroneutral molecule: electroneutrality must be restored, a constraint introducing the $\sum_k \sum_l a_{kl}\Delta q_k$ term. The desciption given in eq. (3.60) is based on one fundamental requirement, the preservation of molecular electroneutrality. The reason for a function F is not far to seek.

In the general context of charge distributions, the role of an atom cannot be described solely by reference to the number of electrons it contributes to the molecular fabric. The centroid of this integrated charge density is equally important. This is where F comes in. The approximations which were introduced partly to gain physical insight and partly to facilitate numerical applications rest on a simplifying assumption, namely, the point-charge model for the saturated hydrocarbons or, more precisely, on the approximate validity of eq. (3.48). These approximations cannot be proven *a priori* to be of minor import. However, the contamination introduced with their use is certainly minimal because they affect only a minor part ($<5\%$) of the dissociation energies which shall be calculated: the final results, i.e., comparisons between calculated and experimental energies, provide an *a posteriori* argument for claiming that these approximations are in fact very good ones. But then, they must be interpreted as general rules which the partitioning of electron densities obtained from accurate electronic wave functions must obey. This is especially important in any extension of an energy theory to cover olefins as well as paraffins; for that extension requires explicit introduction of the $\sigma-\pi$ separation. As a consequence, the point-charge approximation must be reformulated as a 'principle' valid for the averages of electrostatic potentials of the σ orbitals of any given atom in a hydrocarbon molecule. This has two distinct practical advantages: it suggests a very simple numerical correction which allows applications to ethylenic hydrocarbons of the same scheme (in terms of local charges) as used for paraffins and assigns that correction an extremely simple physical interpretation (that of a π centroid displacement) amenable to numerical evaluation from accurate ab initio computations. On the other hand, it involves three important conceptual features: i) the notion of *in situ* atomic orbitals, which must now be introduced explicitly, ii) the postulate that there is a Mulliken-type partitioning of the molecular electron density into atomic orbital contributions which obey the 'principle of orbital balancing', and iii) the implication that a CH bond or a CC bond in an olefin have the same basic properties as in paraffins, except for the effective net charges of the atoms involved. The remarkably effective orbital balancing decribed here signals a general propensity of the local electronic structures to resist modifications and suggests the idea that atomic charge variations should perhaps also be regarded as events occurring most reluctantly.

Coming now to conjugation, results like those of Malrieu *et al.* play an important role in the discussion of polyunsaturated systems. They offer the theoretical background for the well-known thermochemical stabilization of a chemical bond due to conjugation. This stabilization is not reflected in the $\sum_k \sum_l a_{kl}\Delta q_k + F$ part of (3.36). It must be included in the definition of the relevant reference bond energy. This formulation, which is the simplest possible one, is adequate because it turns out that, for all practical purposes, the amount of conjugation associated with a given bond can be treated as a constant (representing a rough but acceptable approximation) unless, of course, changes of molecular structure force a disruption of conjugation. Usually, this causes no difficulties in applications of (3.36), but requires some advance knowledge about molecular structure.

Bibliography

[1] T. Berlin, *J. Chem. Phys.*, **19**, 208 (1951).

[2] K. Ruedenberg, *in* "Localization and Delocalization in Quantum Chemistry", vol. I, p 223, Reidel, Dordrecht, 1975.

[3] W. Kutzelnigg, *in* "The Concept of the Chemical Bond", Z. B. Maksić (ed.), Part 2, p 1, Springer-Verlag, Berlin, 1990.

[4] S. Fliszár, *J. Am. Chem. Soc.*, **102**, 6946 (1980).

[5] G. Del Re, *Gazz. Chim. Ital.*, **102**, 929 (1972).

[6] S. Fliszár, "Charge Distributions and Chemical Effects", Springer-Verlag, New York, 1983.

[7] S. Fliszár, M. Foucrault, M.-T. Béraldin, and J. Bridet, *Can. J. Chem.*, **59**, 1074 (1981).

[8] J. R. Platt, *Handb. Phys.*, **371c**, 188 (1961).

[9] R. F. W. Bader, "Atoms in Molecules, a Quantum Theory", Clarendon Press, Oxford, 1990.

[10] G. Del Re, P. Otto, and J. Ladik, *Isr. J. Chem.*, **19**, 265 (1980).

[11] Z. B. Maksić and K. Rupnik, *Croat. Chem. Acta*, **56**, 461 (1983).

[12] P.-O. Löwdin, *J. Chem. Phys.*, **18**, 365 (1950); *ibid.*, **21**, 374 (1953); E. R. Davidson, *J. Chem. Phys.*, **46**, 3320 (1967); K. Jug, *Theor. Chim. Acta*, **31**, 63 (1973); *ibid.*, **39**, 301 (1975).

[13] R. G. Parr, *Int. J. Quant. Chem.*, **26**, 687 (1984).

[14] G. Del Re and C. Barbier, *Croat. Chem. Acta*, **57**, 787 (1984).

[15] R. S. Mulliken, *J. Am. Chem. Soc.*, **88**, 1849 (1966).

[16] G. Del Re, *Adv. Quantum Chem.*, **8**, 95 (1974).

[17] S. Fliszár, G. Del Re, and M. Comeau, *Can. J. Chem.*, **63**, 3551 (1985).

[18] S. Fliszár, G. Cardinal, and N. A. Baykara, *Can. J. Chem.*, **64**, 404 (1986).

[19] S. Fliszár, G. Cardinal, and M.-T. Béraldin, *J. Am. Chem. Soc.*, **104**, 5287 (1982).

[20] A. R. H. Cole, G. M. Mohay, and G. A. Osborne, *Spectrochim. Acta*, **A23**, 909 (1967); D. J. Marais, N. Sheppard, and B. P. Stoicheff, *Tetrahedron*, **17**, 163 (1962); A. Almenningen, O. Bastiansen, and M. Traetteberg, *Acta Chem. Scand.*, **12**, 1221 (1958); V. Schomaker and L. Pauling, *J. Am. Chem. Soc.*, **61**, 1769 (1939).

[21] W. J. Hehre and J. A. Pople, *J. Am. Chem. Soc.*, **97**, 6941 (1975); P.N. Skancke and J. E. Boggs, *J. Mol. Struc.*, **16**, 179 (1973); B. Dunbacher, *Theoret. Chim. Acta*, **23**, 346 (1972); U. Pincelli, B. Caldoli, and B. Levy, *Chem. Phys. Lett.*, **13**, 249 (1972); L. Radom and J. A. Pople, *J. Am. Chem. Soc.*, **92**, 4786 (1970).

[22] K. Kuchitsu, T. Fukuyama, and Y. Morino, *J. Mol. Struc.*, **1**, 463 (1967); W. Haugen and M. Traetteberg, *Acta Chem. Scand.*, **20**, 1726 (1966).

[23] C. A. Coulson, *Proc. R. Soc.*, **A169**, 413 (1939).

[24] R. S. Mulliken, *Tetrahedron*, **6**, 68 (1959).

[25] S. Skaarup, J. E. Boggs, and P. N. Skancke, *Tetrahedron*, **32**, 1179 (1976); B. M. Mikhailov, *Tetrahedron*, **21**, 1277 (1965); H. J. Bernstein, *Trans. Faraday Soc.*, **67**, 1649 (1961); H. C. Longuet–Higgins and L. Salem, *Proc. R. Soc.*, **A251**, 172 (1959).

[26] H. Kollmar, *J. Am. Chem. Soc.*, **101**, 4832 (1979).

[27] J. P. Daudey, G. Trinquier, J. C. Barthelat, and J.-P. Malrieu, *Tetrahedron*, **36**, 3399 (1980).

[28] G. W. Wheland, "Resonance in Organic Chemistry", Wiley, New York, 1955.

Chapter 4

Bond Dissociation Energies

4.1 Scope

Consider a polyatomic molecule and focus attention on a particular atom pair, k and l, forming a bond with intrinsic energy ε_{kl}. This energy cannot be observed in isolation. What can be measured (in principle) is D_{kl}, the bond dissociation energy, i.e., the energy required for the physical breaking of that bond. Now, D_{kl} depends on a number of events accompanying bond breaking, e.g., possible geometry and hybridization changes affecting the fragments. Briefly, $D_{kl} \neq \varepsilon_{kl}$ in polyatomic molecules[1]. It is understood that D_{kl}, like ε_{kl}, refers to molecules at their potential minimum.

Intrinsic bond energies and bond dissociation energies meet different practical needs. The former play an important role in the description of ground-state molecules. Dissociation energies come into play when molecules undergo reactions. Now, any interaction between a molecule and its environment (such as complex formation or adsorption onto a metallic surface, for example) affects its electron distribution and thus the energies of its chemical bonds. This part is handled by eq. (3.39). If we figure out the relationship between dissociation and intrinsic bond energies, we could begin to unterstand how the environment of a molecule can promote or retard the dissociation of one or another bond of particular interest in that molecule.This

---

[1] For diatomic molecules, of course, $\Delta E_a^{\bullet} = \varepsilon_{kl} = D_{kl}$. In polyatomic molecules, consideration of a step-wise cleavage of all the bonds gives $\Delta E_a^{\bullet} = \sum_{k<l} D_{kl}$, which does not imply that the individual dissociation energies are the same as the intrinsic bond energies of the original molecule. In methane, for example, $\varepsilon_{\mathrm{CH}} = 104.86$ kcal mol^{-1}, but the dissociation energy accompanying the cleavage $\mathrm{CH_4} \rightarrow \mathrm{CH_3\cdot} + \mathrm{H}$ is $D_{\mathrm{CH}} = 111.4$ kcal mol^{-1}.

general outlook hints at a rich potential of future research exploiting charge analyses to gain insight into bond energies, first, and, going from there, into matters of great import regarding the dissociation of chemical bonds. Ours is just a peep in that direction.

It takes little time to demonstrate the relationship between D_{kl} and ε_{kl}. The theory [1][2] is a logical extension of Chapter 3. This is why we discuss it now. The result, however, is best understood with the help of examples. The required numerical material is found in Chapters 5 and 6, but we use it here to avoid disrupting the description of dissociation energies.

4.2   Theory

Consider a molecule KL and its dissociation into two fragments, K· and L·, that is

$$KL \rightarrow K \cdot + L \cdot$$

At this point we still regard the molecule as a whole, at its potential minimum, but this does not prevent us from recognizing the two subunits, K and L, which are eventually going to separate. Let k be the atom found in the subunit K and l the atom of L that form the k–l bond undergoing cleavage. Suppose we know ε_{kl}, i.e., the intrinsic energy of the k–l bond in the original ground-state molecule. We want to calculate D_{kl} or, more precisely, determine how D_{kl} and ε_{kl} are related to one another.

The exact definition of D_{kl} is

$$D_{kl} = \Delta E_a^*(KL) - \Delta E_a^*(K\cdot) - \Delta E_a^*(L\cdot) \tag{4.1}$$

where $\Delta E_a^*(KL)$ represents the atomization energy of the ground-state molecule KL. Similarly, $\Delta E_a^*(K\cdot)$ and $\Delta E_a^*(L\cdot)$ are the atomization energies of the ground-state free radicals K· and L·, respectively. (One or both reaction products could be atoms, depending on what KL is.)

Now examine the bond-by-bond partitioning of $\Delta E_a^*(KL)$ indicated by eq. (3.5), but rename the bonds as follows

$$\Delta E_a^* = \sum_{t<v} \varepsilon_{tv} - E_{nb} \tag{4.2}$$

to avoid confusion with the k–l bond under scrutiny. When we calculate $\Delta E_a^*(KL)$, the summation in (4.2) obviously collects all existing bonds, including the k–l bond. Here we focus attention on this k–l bond. So we subdivide the right-hand side of eq. (4.2) and consider only the bonds found in the molecular subunit K and the nonbonded interactions involving only the atoms belonging to K. The sum of all these contributions is $\Delta E_a^*(K)$. This is the part of $\Delta E_a^*(KL)$ that is associated with the group of atoms of K as it exists in that particular molecule KL prior to its dissociation. We proceed similarly with the subunit L and obtain $\Delta E_a^*(L)$. The sum $\Delta E_a^*(K)$ + $\Delta E_a^*(L)$ thus collects the intrinsic energies of all the bonds found in the original molecule except one, ε_{kl}, and all the nonbonded interactions, except those between the atoms of the subunit K and those of the subunit L, represented by $E_{nb}(K \cdots L)$. The energy balance satisfying exactly eq. (4.2) for the original molecule KL is therefore

$$\Delta E_a^*(KL) = \Delta E_a^*(K) + \Delta E_a^*(L) + \varepsilon_{kl} - E_{nb}(K \cdots L). \qquad (4.3)$$

When K and L are identical in the molecule (as in CH_3–CH_3, for example, but not in $HOH \cdots OH_2$), each of these groups is necessarily electroneutral. Electroneutral subunits in a molecule, taken exactly as they are in molecules like KK, LL, etc., are identified by the superscript "o". Under these conditions, it follows from eq. (4.3) that

$$\Delta E_a^*(K^\circ) = \frac{1}{2} [\Delta E_a^*(KK) - (\varepsilon_{kl} - E_{nb}(K \cdots K))] \qquad (4.4)$$

which is useful in numerical applications.

The energy formulas (4.1) and (4.3) translate straightforward applications of first-principle energy conservation. No hypothesis is involved other than that related to the assumed validity of separating the molecular binding energy into bonded and nonbonded contributions, eq. (4.2). Equation (4.1) features the dissociation energy, D_{kl}, and eq. (4.3) the corresponding intrinsic bond energy, ε_{kl}. At last we can examine how these bond terms are related to one another.

The key argument is rooted in a simple observation concerning any dissociation KL→ K· + L·. *The individual subunits K and L are in general not electroneutral while they are part of the original host molecule whereas the corresponding radicals certainly satisfy electroneutrality.* A charge neutralization accompanies the transformations K→K· and L→L·. This constraint solves our problem. A few examples[2] (Table 4.1) help understanding why

---

[2] The $\Delta E_a^*(K)$ energies were derived from eq. (4.2) by means of bond-by-bond calculations, eq. (3.39), and Del Re's approximation, eq. (3,11), for the nonbonded contributions [2]. Atomization energies of the host molecules, calculated in precisely the same manner, agree within ~0.2 kcal mol^{-1} (average deviation) with their experimental counterparts.

TABLE 4.1. Net charges (me) and $\Delta E_a^*(K)$
energies (kcal mol^{-1}) of methyl, ethyl and
*iso*propyl groups in selected host molecules

K	Host molecule	Charge	$\Delta E_a^*(K)$
CH$_3$	CH$_4$	9.05	314.53
	CH$_3$COCH$_3$	3.60	318.45
	CH$_3$CH$_3$	0.00	320.34
	CH$_3$C$_2$H$_5$	−2.65	322.15
	CH$_3$*i*-C$_3$H$_7$	−5.02	323.80
	CH$_3$OCH$_3$	−10.47	327.50
C$_2$H$_5$	C$_2$H$_5$COC$_2$H$_5$	33.60	591.50
	C$_2$H$_5$CH$_3$	2.65	610.81
	C$_2$H$_5$C$_2$H$_5$	0.00	612.78
	C$_2$H$_5$OC$_2$H$_5$	−2.59	615.09
i-C$_3$H$_7$	*i*-C$_3$H$_7$CO*i*-C$_3$H$_7$	50.30	875.00
	i-C$_3$H$_7$CH$_3$	5.02	903.63
	i-C$_3$H$_7$O*i*-C$_3$H$_7$	3.55	904.50
	i-C$_3$H$_7$*i*-C$_3$H$_7$	0.00	906.67

this argument is important. These examples indicate *i*) the net charge of
K, i.e., to what extent K departs from exact electroneutrality as long as it
is part of its host molecule and *ii*) the corresponding response in $\Delta E_a^*(K)$
energy, i.e., how $\Delta E_a^*(K)$ 'feels' this departure from electroneutrality. This
response is significant, even for relatively small departures from electroneutrality. A fragment K increases its thermochemical stability when it gains
electronic charge from its environment: $\Delta E_a^*(K)$ becomes larger.

Let us now apply what we have learned from these results and develop
our strategy with the help of the examples worked out for the methyl groups.
These groups differ from one another in $\Delta E_a^*(K)$ energy and we know by how
much. Therefore, it suffices to know how anyone of them differs from the
true CH$_3$· radical in order to gain the same information for the other CH$_3$
groups. A convenient reference is the neutral CH$_3$ of ethane because it is
isoelectronic with CH$_3$·. Its $\Delta E_a^*(K)$ value is deduced from (4.4) and is
written $\Delta E_a^*(K°)$. This choice is convenient because it permits the association of energy changes $\Delta E_a^*(K) - \Delta E_a^*(K°)$ with charge neutralization—
hence the term CNE = charge neutralization energy. The methyl group of
CH$_4$, for example, is electron deficient by 9.05 me. When recovering this
charge during the cleavage CH$_4 \rightarrow$ CH$_3$· + H, its energy *decreases* by CNE

$= -5.81$ kcal mol^{-1} meaning that $\Delta E_a^*(CH_3)$ *increases* by that amount. Similarly, when the CH_3 of propane restores its electroneutrality by losing its excess electronic charge, -2.65 me, its energy increases by CNE $= 1.81$ kcal mol^{-1}. Note that these CNE results include all possible effects due to geometry changes because they are derived from $\Delta E_a^*(K)$ energies satisfying eqs. (4.2)–(4.4) which hold for molecules at equilibrium geometry with the proper charges corresponding to this situation.

The generalization of what we have just learned is straightforward. CNE is a useful concept which permits us to relate any K in a molecule, described by $\Delta E_a^*(K)$, to the corresponding electroneutral K°, described by $\Delta E_a^*(K^\circ)$. In order to learn how any K embedded in its host molecule differs in energy from the ground-state radical K·, it suffices to know once and for all how K° differs from K·. This is the so-called reorganizational energy

$$RE = \Delta E_a^*(K^\circ) - \Delta E_a^*(K\cdot) \tag{4.5}$$

which reflects all possible geometry and hybridization changes accompanied, of course, by significant redistributions of electronic charge. The selection of $\Delta E_a^*(K^\circ)$ as a reference is only one of the possible choices. It is arbitrary but convenient because it makes sense to first get the number of electrons right, then relax everything else. Having now secured this idea, there is an easy way of exploiting it.

Two radicals are formed in any dissociation KL→K· + L·. For the problem at hand, it is more convenient to consider them jointly rather than proceeding *via* calculations of the individual CNE contributions which are $\Delta E_a^*(K) - \Delta E_a^*(K^\circ)$ on the one hand and $\Delta E_a^*(L) - \Delta E_a^*(L^\circ)$, on the other. The sum $\Delta E_a^*(K) + \Delta E_a^*(L)$ differs in principle from that of the corresponding electroneutral fragments, $\Delta E_a^*(K^\circ) + \Delta E_a^*(L^\circ)$. The difference between these sums is CNE, that is

$$CNE = [\Delta E_a^*(K) - \Delta E_a^*(K^\circ)] + [\Delta E_a^*(L) - \Delta E_a^*(L^\circ)]. \tag{4.6}$$

Using now eq. (4.3), it follows that

$$CNE = \Delta E_a^*(KL) - \varepsilon_{kl} + E_{nb}(K\cdot\cdot L) - \Delta E_a^*(K^\circ) - \Delta E_a^*(L^\circ). \tag{4.7}$$

Equation (4.7) translates the general idea that molecular subunits which are not individually electroneutral in the host molecule must restore their correct numbers of electrons when dissociation takes place.

Equation (4.7) is important in another way: it is the key for the description of bond dissociation energies. Indeed, using the definition of D_{kl}, eq. (4.1), and that of reorganizational energy, equation (4.5), it follows from (4.7) that

$$D_{kl} = \varepsilon_{kl} - E_{\mathrm{nb}}(\mathrm{K} \!\cdot\!\! \cdot \mathrm{L}) + \mathrm{CNE} + \mathrm{RE}(\mathrm{K}) + \mathrm{RE}(\mathrm{L}). \qquad (4.8)$$

This energy formula [1][2] is general and suffers from no approximations in that all the appropriate bonded and nonbonded contributions are formally taken care of and because the electroneutrality requirements are met in the definition of reorganizational energy. Only dissociations yielding electroneutral products need be considered because the formation of ions, $\mathrm{KL} \rightarrow \mathrm{K}^+ + \mathrm{L}^-$, can be treated along the same lines, with final corrections involving the ionization potential of K and the electron affinity of L.

The formula for D_{kl} contains energy terms, namely, ε_{kl} and $E_{\mathrm{nb}}(\mathrm{K}\!\cdot\!\!\cdot\mathrm{L})$, which depend only on the properties of the reactant KL. The calculation of CNE, however, requires additional information for use in eq. (4.4). Finally, this formula also contains information about the products, which is included in the reorganizational-energy terms. In other words, there is no way dissociation energies could be predicted exclusively in terms of the reactant's ground-state properties. Though unfortunate, this point must be clear in our minds. Equation (4.8) is only capable of telling us why D_{kl} and ε_{kl} are different. This is nevertheless instructive, because all the terms appearing in (4.8) are understood on physical grounds. The true merits of this equation are revealed by a survey of the leading terms governing bond dissociations when the cleavage occurs i) in the 'interior' of a molecule, i.e., when both K and L are polyatomic groups, ii) in its peripheral region, i.e., when K (or L) is an atom, or iii) when it concerns 'exterior' bonds formed by a molecule, such as hydrogen bonds.

The application of (4.8) to diatomic molecules merits a few words, to help develop familiarity with this equation. Evidently, $E_{\mathrm{nb}} = 0$ for these molecules. Equation (4.5) gives $\mathrm{RE}(\mathrm{K}) = \Delta E_{\mathrm{a}}^*(\mathrm{K}^\circ)$ because $\Delta E_{\mathrm{a}}^*(\mathrm{K}\cdot) = \Delta E_{\mathrm{a}}^*(\mathrm{atom\ K}) = 0$. Eq. (4.2) becomes $\Delta E_{\mathrm{a}}^*(\mathrm{KL}) = \varepsilon_{kl}$. Using it in eq. (4.7) we get $\mathrm{CNE} + \mathrm{RE}(\mathrm{K}) + \mathrm{RE}(\mathrm{L}) = 0$ for any diatomic molecule. Finally, eq. (4.8) yields the well-known result $D_{kl} = \varepsilon_{kl}$. On the other hand, eq. (4.4) shows that $\Delta E_{\mathrm{a}}^*(\mathrm{K}^\circ) = 0$, meaning that $\mathrm{RE}(\mathrm{K}) = 0$ when K is an atom. This means that $\mathrm{CNE} = 0$ for diatomic molecules. Now, CNE accounts for the fact that any charge imbalance affects the chemical bonds and the nonbonded terms making up the energies of the subunits K and L. For that reason it concerns only polyatomic subunits. In writing CNE $= 0$ for heteronuclear diatomic molecules, it is understood that the charge neutralization accompanying their cleavage is part of ε_{kl}.

TABLE 4.2. The calculation of reorganizational energies, kcal mol^{-1}

	K	$\Delta E_a^*(KK)$	ε_{kl}	$\Delta E_a^*(K°)$	$\Delta E_a^*(K\cdot)$	RE
1	CH_3	710.54	69,63	320.34	307.89	12.45
2	C_2H_5	1298.10	72.38	612.78	602.73	10.05
3	$n\text{-}C_3H_7$	1885.69	73.39	906.07	896.88	9.19
4	$i\text{-}C_3H_7$	1887.16	73.71	906.67	899.68	6.99
5	$n\text{-}C_4H_9$	2473.21	73.06	1199.99	1191.05	8.94
6	$i\text{-}C_4H_9$	2476.03	74.14	1200.88	1193.26	7.62
7	$s\text{-}C_4H_9$	2473.55	74.44	1199.50	1194.85	4.65
8	$t\text{-}C_4H_9$	2477.09	73.87	1201.61	1199.16	2.45
9	$c\text{-}C_6H_{11}$	3401.42	75.27	1663.03	1657.22	5.81

4.3 Applications

The hydrocarbons are good candidates to bring out the merits of eq. (4.8), its salient features and some intricacies plaguing the interpretation of trends observed for the breaking of chemical bonds.

Let us first get the required reorganizational energies. Those indicated in Table 4.2 were calculated [2] from (4.5) using theoretical $\Delta E_a^*(K°)$ results given by eq. (4.4) and selected atomization energies for the free radicals[3], with proper consideration of the nonbonded terms[4].

Now we come to the interesting part: the carbon–carbon bonds represent typical examples of 'interior bonds'.

The carbon skeleton of alkanes

The first thing to do is to calculate CNE by means of equation (4.7). The appropriate theoretical $\Delta E_a^*(KL)$ results are indicated in Table 4.3, along with the corresponding CC bond energies. The nonbonded terms are from [2] and the $\Delta E_a^*(K°)$'s are those of Table 4.2. The CNE results deduced in

---

[3] The $\Delta E_a^*(K\cdot)$ values are reported with a precision that is not warranted by our actual knowledge. They are used as indicated primarily in order to facilitate recalculations without being continually bothered by rounding-off errors. None of the conclusions reached here critically depends on the assumed precision of these energies.

[4] The latter were calculated according to Del Re, eq. (3.11), using the charges described in Chapter 5. When K and L are alkyl groups, $E_{nb}(K\cdot L)$ is of the order of ~ -0.2 kcal mol^{-1} and much smaller, but positive, when L = hydrogen. Detailed results are tabulated in [2].

TABLE 4.3. Charge neutralization and dissociation energies of the CC bonds of selected saturated hydrocarbons, kcal mol^{-1}

Bond	ΔE_a^*(KL)	ε_{CC}	CNE	D_{CC} Exptl.	D_{CC} Eq.(4.9)
CH_3-CH_3	710.54	69.63	0.00	94.76	94.9
$CH_3-C_2H_5$	1004.29	71.14	−0.16	03.67	94.0
$CH_3-n-C_3H_7$	1298.10	71.56	−0.06	93.33	93.5
$CH_3-i-C_3H_7$	1300.02	72.42	0.42	92.45	92.2
$CH_3-n-C_4H_9$	1591.90	71.41	−0.04	92.96	93.1
$CH_3-i-C_4H_9$	1593.17	71.91	−0.13	92.02	92.3
$CH_3-s-C_4H_9$	1593.17	72.53	0.63	90.43	90.0
$CH_3-t-C_4H_9$	1596.06	73.14	0.81	89.01	88.4
$C_2H_5-C_2H_5$	1298.10	72.38	0.00	92.64	92.8
$C_2H_5-n-C_3H_7$	1591.90	72.89	0.00	92.29	92.5
$C_2H_5-i-C_3H_7$	1593.17	73.23	0.36	90.76	90.6
$C_2H_5-n-C_4H_9$	1885.69	72.74	0.01	91.91	92.1
$C_2H_5-i-C_4H_9$	1887.00	73.23	−0.03	91.01	91.2
$C_2H_5-s-C_4H_9$	1886.33	73.55	0.36	88.75	88.6
$C_2H_5-t-C_4H_9$	1888.41	73.63	0.31	86.52	86.5
$n-C_3H_7-i-C_3H_7$	1887.00	73.81	0.32	90.44	90.3
$n-C_3H_7-s-C_4H_9$	2180.15	74.10	0.34	88.42	88.3
$i-C_3H_7-i-C_3H_7$	1887.16	73.71	0.00	87.80	88.0
$i-C_3H_7-s-C_4H_9$	2180.32	74.05	−0.01	85.79	86.0
$i-C_3H_7-t-C_4H_9$	2182.36	73.88	0.15	83.52	83.7
$i-C_4H_9-s-C_4H_9$	2475.21	74.49	0.22	87.10	87.1
$s-C_4H_9-s-C_4H_9$	2473.55	74.44	0.00	83.85	84.1
$s-C_4H_9-t-C_4H_9$	2476.04	74.47	0.40	82.03	81.9
$t-C_4H_9-t-C_4H_9$	2477.09	73.87	0.00	78.77	79.1

this manner are displayed in Table 4.3.

The important point is that the CNE contributions are small in comparison with the significant changes of the ε_{kl} energies. This was to be expected because charge neutralization between alkyl groups benefits from important compensations. Citing propane as an example, each of its methyl groups carries an excess negative charge, −2.65 me. This charge is necessarily given back to the ethyl moiety when a CC bond dissociates, thus reducing the ΔE_a^*(K) energy of the CH_3 part from 322.15 to 320.34 kcal mol^{-1} and increasing that of C_2H_5 from 610.81 to 612.78 kcal mol^{-1}. (Check with Table 4.1.) The combined effect is a small change, −0.16 kcal mol^{-1}, in total CNE energy. Nonzero CNE's should be generally expected when K≠L, but their smallness and the almost negligible variations of the nonbonded contributions [2] suggest that $CNE - E_{nb}(K \cdot \cdot L) \simeq$ constant should represent a valid approximation, that is

$$D_{CC} \simeq \varepsilon_{CC} + RE(K) + RE(L) + 0.33 \text{ kcal mol}^{-1} \qquad (4.9)$$

where the constant (0.33) is the average of the CNE and nonbonded contributions evaluated for the thirtysix molecules constructed from the eight alkyl groups considered here. The associated error (standard deviation = 0.26 kcal mol^{-1}) is certainly acceptable[5].

The approximate validity of eq. (4.9) highlights the leading role played by the reorganizational energies but also demonstrates the importance of using the correct intrinsic bond energies, whose values range from 69.63 to $\sim$74.5 kcal mol^{-1} in these examples. This brings us to examine the formal resemblance and the conceptual difference between (4.9) and Sanderson's approximation [3]. The latter can be deduced from eqs. (4.1) and (4.3) by redefining the reorganizational energies as $\Delta E_a^*(K) - \Delta E_a^*(K\cdot)$—an approach oblivious of the fact that K and K$\cdot$ are not isoelectronic in most cases, but approximately valid as long as the CNE terms remain sufficiently small. The point is that Sanderson's formula, $D_{CC} = \varepsilon_{CC} + RE(K) + RE(L)$, treats ε_{CC} as a constant. This simplifying hypothesis did not appear as a conceptual stumbling-block at that time—and, indeed, many interesting results were obtained in this fashion—but, as we now know, any scheme postulating constant bond energies conflicts in principle with the basics demanding conservation of molecular electroneutrality. The quality of the theoretical intrinsic bond energies is clearly reflected in the quality of the bond dissociation energies given by eq. (4.9).

The carbon–hydrogen bonds

The carbon–hydrogen bonds are the prototype of 'peripheral bonds'. In the perspective of the CNE effect, the extraction of an atom that was partially charged in the host molecule affects only the bond energies of the polyatomic fragment left behind because the extracted atom forms no bonds and cannot compensate for the energy changes induced in the other fragment. Briefly, any cleavage of this sort is expected to involve a significant charge neutralization energy. CH bonds fall in with this description.

---

[5]This conclusion also holds for dissociation enthalpies $DH(KL) = \Delta H_f(K\cdot) + \Delta H_f(L\cdot) - \Delta H_f(KL)$, where the ΔH_f's are the appropriate enthalpies of formation. This is due to structure-related regularities characterizing zero-point and heat-content energies (Appendix A). Eq. (4.9) thus gives [2] $DH(CC) \simeq \varepsilon_{CC} + RE(K) + RE(L) - 5.9 \text{ kcal mol}^{-1}$.

TABLE 4.4. Dissociation and
intrinsic energies of selected
CH bonds, kcal mol^{-1}

CH Bond	D_{CH}	ε_{CH}
CH_3-H	111.42	104.86
C_2H_5-H	107.81	106.81
$n\text{-}C_3H_7-H$	107.41	107.13
$i\text{-}C_3H_7-H$	104.61	108.72
$n\text{-}C_4H_9-H$	107.05	107.23
$i\text{-}C_4H_9-H$	106.76	107.59
$s\text{-}C_4H_9-H$	103.25	109.35
$t\text{-}C_4H_9-H$	100.86	110.89

A few examples suffice to get us at the heart of the problem. Examine typical alkane CH bonds (Table 4.4). The comparison between their dissociation energies and their intrinsic bond energies is intriguing, to say the least. It suggests (Figure 4.1) that those bonds which contribute more to thermochemical stability are the ones that break up more easily. This is certainly a point requiring clarification. In addition, Sanderson's claim [3] that all contributing CH bond energies are equal should be examined. On the one hand, this claim is diametrically opposed in conceptual content to the basics of our description of bond energies—the preservation of molecular electroneutrality—still, on the other hand, it deserves more than passing attention because of the undisputed success of Sanderson's approach.

FIGURE 4.1. A comparison between CH dissociation and intrinsic bond energies, kcal mol^{-1}.

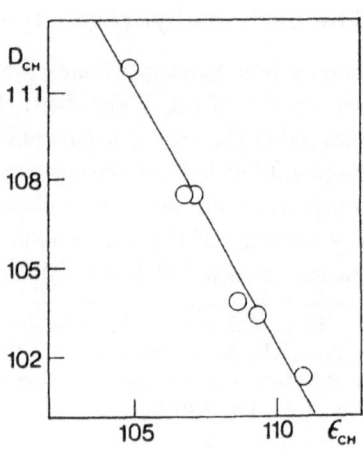

TABLE 4.5. CNE energies, kcal mol^{-1}, for C(sp^3)-H
and C(sp^2)-H bond dissociations

Bond	ε_{CH}	CNE	ε_{CH} + CNE
CH$_3$-H	104.86	-5.81	99.05
C$_2$H$_5$-H	106.81	-9.01	97.80
n-C$_3$H$_7$-H	107.13	-8.87	98.26
i-C$_3$H$_7$-H	108.72	-11.08	97.64
n-C$_4$H$_9$-H	107.23	-9.08	98.15
i-C$_4$H$_9$-H	107.59	-8.42	99.17
s-C$_4$H$_9$-H	109.35	-10.74	98.61
t-C$_4$H$_9$-H	110.89	-10.50	98.39
c-C$_6$H$_{11}$-H	109.28	-11.50	97.78
CH$_2$:CHCH$_2$-H	107.3	-8.6	98.69
CH$_2$:CHCH$_2$CH$_2$-H	107.4	-8.9	98.52
C$_6$H$_5$CH$_2$-H	108.0	-9.0	99.0
CH$_2$:CH-H	105.72	-3.11	102.61
CH$_3$CH:CH-H	106.0	-3.9	102.1
(CH$_3$)$_2$C:CH-H	104.6	-1.3	103.3
CH$_2$:C(CH$_3$)-H	109.9	-7.1	102.8
C$_6$H$_5$-H	112.36	-9.79	102.57

It turns out that the CNE part accompanying CH dissociations is not only quantitatively important, but also produces a remarkable effect (Table 4.5). Indeed, CNE compensates the major part, if not all, of the differences existing between ε_{CH} energies: ε_{CH} + CNE is nearly constant for all the C(sp^3)-H bonds, on the one hand, and also for the C(sp^2)-H bonds, on the other. Hence, incorporating now the small (< 0.05 kcal mol^{-1}) nonbonded interactions into this approximation, we rewrite eq. (4.8) as follows

$$D_{CH} \simeq \text{constant} + \text{RE(K)} \qquad (4.10)$$

where the constant, 98.40 or 102.65 kcal mol^{-1} for C(sp^3)-H and C(sp^2)-H bonds, respectively, stands for the appropriate average value of ε_{CH} + CNE $- E_{nb}(K \cdot \cdot H)$.

The ε_{CH} energies examined here range from $\sim$104.9 to $\sim$110.9 and from $\sim$105.7 to $\sim$113.5 kcal mol^{-1} for C(sp^3)-H and C(sp^2)-H bonds, respectively (Table 4.5). These are the genuine bond energies of eq. (3.39). CNE disguises them in such a way that, when viewed from the perspective of dissociation energies, all CH bonds involving the same type of carbon are perceived as if they were equal in energy, to a good approximation. This result, which is typical for peripheral bonds, explains the nature of 'constant CH bond energies' and vindicates Sanderson's claim concerning their use in problems of bond dissociation along the lines indicated by eq. (4.10). Finally, the reason for the existence of a correlation between D_{CH} and ε_{CH} energies (Figure 4.1) is given by equation (4.10) and the approximate linear decrease of reorganizational energies for increasingly larger ε_{CH} energies (Figure 4.2).

4.4   Generalization

Non-negligible CNE contributions typify the dissociation of peripheral bonds, K–X→K· + atom X. The larger the net charge on X, the larger is |CNE|. As regards the intrinsic properties of these bonds, they depend, of course, on the type of bond-forming atoms. They also depend on their charges. The remarkable thing with CH bonds is that CNE compensates for these charge effects when it comes down to breaking them, at least to a good approximation, eq. (4.10). Carbon–halogen bonds follow a similar pattern [1] (Table 4.6) but, of course, this does not mean that all types of peripheral bonds are represented by a simple approximation like (4.10).

The main characteristic of the 'interior' bonds is that their dissociation involves in principle minor CNE contributions. The carbon skeleton of the alkanes offers a typical example, epitomized in the approximation (4.9). The reorganizational energies are, of course largely responsible for the observed dissociation energies but the role of the intrinsic bond energies now begins to surface. This paves the way to novel applications.

Suppose we study a bond in the interior of an isolated molecule, ε_{kl}^{o} being the intrinsic energy of that bond and D_{kl}^{o} its dissociation energy. Then we

FIGURE 4.2. A comparison between reorganizational energies and theoretical bond energies of alkane CH bonds, in kcal mol^{-1}. The numbering refers to that given in Table 4.2.

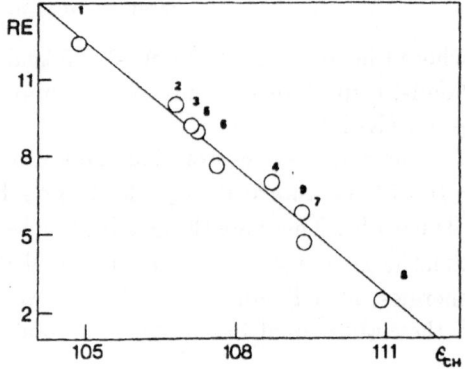

TABLE 4.6. Dissociation and ε_{CX} +
CNE $-E_{nb}$ energies of haloalkanes,
kcal mol^{-1}

K	X	ε_{CX}+CNE $-E_{nb}$	D_{CX}
CH_3	Cl	74.5	86.9
C_2H_5	Cl	76.0	86.1
C_3H_7	Cl	76.2	85.4
C_4H_9	Cl	75.3	84.2
CH_3	Br	60.4	72.8
C_2H_5	Br	62.5	72.5
C_3H_7	Br	63.0	72.2
C_4H_9	Br	63.1	72.0
CH_3	I	46.3	58.7
C_2H_5	I	47.7	57.8
C_3H_7	I	48.1	57.3

study the same molecule but let it interact with its environment, say, an other molecule, or a metallic surface, etc. The external perturbation modifies the wave function and thus the local electron densities at atoms k and l. The intrinsic energy of the k–l bond is now

$$\varepsilon_{kl} = \varepsilon_{kl}^{\circ} + \Delta\varepsilon_{kl}.$$

Whether we break this bond in the isolated or in the perturbed molecule, we end up with the same products, meaning that the reorganizational energies are the same in both cases. Any possible change affecting CNE can be expected to be minor since it is usually a small quantity to begin with. A reasonable approximation of (4.9) describing this situation is thus

$$\Delta D_{kl} \approx \Delta\varepsilon_{kl} \qquad (4.11)$$

where $\Delta D_{kl} = D_{kl} - D_{kl}^{\circ}$. Using eq. (3.39) we obtain [4]

$$\Delta D_{kl} \approx a_{kl}\Delta q_k + a_{lk}\Delta q_l \qquad (4.12)$$

where Δq_k and Δq_l indicate how the charges of atoms k and l, respectively, differ, in the perturbed molecule, from those of the isolated molecule.

The loss of 1 millielectron (i.e., $\Delta q = +1$ me) has a significant repercussion on D_{kl}. The a_{kl} results of Table 4.7 indicate, for example, that $\Delta q = 1$ me at a hydrogen atom lowers the energy of a CH bond by 0.632 kcal mol^{-1}.

TABLE 4.7. Selected a_{kl} values for single bonds and the
bond weakening due to the loss of 1 me at both atoms

Bond	a_{kl}, kcal mol^{-1}me^{-1}		Bond weakening
HC	$a_{HC} = -0.632$	$a_{CH} = -0.247$	0.879
HN	$a_{HN} = -0.814$	$a_{NH} = -0.200$	1.014
HO	$a_{HO} = -1.013$	$a_{OH} = -0.229$	1.242
CC	$a_{CC} = -0.488$		0.976
CN	$a_{CN} = -0.593$	$a_{NC} = -0.441$	1.034
CO	$a_{CO} = -0.712$	$a_{OC} = -0.505$	1.217

Table 4.7 suggests the following rule of thumb: *the loss of 1 millielectron
at each atom forming a single bond translates into a bond weakening of ~ 1
kcal mol^{-1}*. Surely, microscopic changes in local charge densities produce
sizeable themochemical effects. Of all the dissociation formulas encountered
here, eq. (4.12) is probably the one that should prove most useful to chemists:
it offers insight into the response of dissociation in the interior of a molecule
to outside stimuli. It is also the easiest one to use[6].

In closing, let us briefly explore 'weak interactions', i.e., processes involving the exterior of a molecule, such as complex formation. When two
ground-state molecules, A and B, associate without loss of their chemical
identities (as in HOH··OH$_2$), some charge transfer is anticipated in AB. The
CNE part accompanying the dissociation AB→A + B is given by eq. (4.6).
The $\Delta E_a^*(K°)$'s are now simply the atomization energies of A and B, with
RE = 0. The linkage beween A and B should certainly not be considered
as a 'normal' chemical bond, but should be more appropriately included in
$E_{nb}(A \cdot\cdot B)$ although it may differ from a 'usual' nonbonded interaction. It
is a matter of taste whether or not we single out any particular interaction,
e.g., that between O and H in the hydrogen bond HOH··OH$_2$. Eq. (4.8) thus
reduces to:

$$D_{A\cdot\cdot B} = CNE - E_{nb}(A \cdot\cdot B) \tag{4.13}$$

Equation (4.13) discriminates between intramolecular (CNE) and intermolec-
ular contributions to $D_{A\cdot\cdot B}$. To learn about them is left to the future.

---

[6]A simple example is offered by the CN bond of nitromethane [5]. The monomer,
compared to its trimer (taken as a model for the crystal), reveals that the C and N net
charges change by $\Delta q_C \simeq -8.7$ and $\Delta q_N \simeq 1.1$ me, respectively, upon 'crystallization'. Eq.
(4.12) and the appropriate a_{kl} parameters of Table 4.7 thus indicate that the crystalline
environment reinforces the CN bond by about 4.7 kcal mol^{-1}.

4.5 Conclusions

Our description of bond dissociation energies bears clear similarities with Sanderson's ideas, namely, that a dissociation energy is given by the intrinsic bond energy ε_{kl} plus the appropriate reorganizational energies. This precludes the prediction of dissociation energies using only the ground-state properties of molecules. The new aspects introduced by the present theory lie i) in the use of theoretical, charge-dependent, bond energies, eq. (3.39), which satisfy the demand for molecular electroneutrality, ii) in the appropriate use of electroneutral references for defining reorganizational energies, and iii) with the unavoidable introduction of a new term, the charge neutralization energy, which accounts for the fact that individual molecular subunits may not be electroneutral while they are part of the host molecule whereas the corresponding radicals certainly are. Explicit consideration of this constraint leads directly to a general energy formula describing bond dissociation.

The numerical examples presented for alkanes reveal the minor role played by Coulomb interactions between nonbonded atoms. In the dissociation of carbon–carbon bonds, the mutual neutralization of charge by alkyl groups results in an important mutual cancellation of the individual energy terms associated with this neutralization. The net effect is small. As a consequence, the cleavage of CC bonds is primarily governed by the intrinsic bond energy plus the reorganizational energies of the radicals. The quality of the theoretical bond energies, eq. (3.39), is well reflected in the quality of the calculated dissociation energies. The important cancellation of charge neutralization terms is characteristic and generally anticipated for dissociations involving 'interior' bonds, i.e., bonds formed by polyatomic groups capable of dissipating the overall effects due to charge neutralization between these groups. This property and the use of theoretical intrinsic bond energies offer challenging perspectives to chemists by making it possible to predict how the environment of a molecule can promote or retard the dissociation of one or another interior bond of particular interest in that molecule—an event that can be monitored by means of eq. (4.12).

In contrast, significant charge neutralization energies are typical of dissociations involving peripheral bonds. In the dissociation of carbon–hydrogen bonds, in particular, charge neutralization energies are entirely responsible for a peculiar perception of 'seemingly constant contributing CH bond energies', leaving the reorganizational energy as the sole term regulating the dissociation of a CH bond. That this is true, at least to a good approximation, is known from Sanderson's work. Theory now reveals that, in fact, it is the sum of the genuine intrinsic bond energy, ε_{CH}, plus the charge neutralization energy which remains nearly constant for CH bonds.

Bond dissociation is but one facet of the complex aspects involved in the decription of molecules interacting with their environment. In that broad area, combinations [6] of quantum mechanical models, such as the SCF Reaction Field [7] and the Polarizable Continuum model of Tomasi *et al.* [8], nowadays provide the most successful approaches. Beyond them, in the particular context of bond dissociation, it is primarily eq. (4.12) that adds to our arsenal of investigative tools and hopefully opens new horizons. It is clear, however, that the usefulness of any of the energy formulas encountered so far heavily depends on the availability of reliable charge analyses.

Bibliography

[1] S. Fliszár and C. Minichino, *J. Phys.*, Colloque C4, **48**, 367 (1987).

[2] S. Fliszár and C. Minichino, *Can. J. Chem.*, **65**, 2495 (1987).

[3] R. T. Sanderson, *J. Org. Chem.*, **47**, 3835 (1982); R. T. Sanderson, "Chemical Bonds and Bond Energy", 2nd Ed., Academic Press, New York, 1976.

[4] S. Fliszár, *in* "Chemistry and Physics of Energetic Materials", S. N. Bulusu (ed.), *NATO ASI Series C*, **309**, 143 (1990), Kluwer Acad. Publ., Dordrecht, NL.

[5] Simone Odiot, M. Blain, E. Vauthier, and S. Fliszár, *J. Mol. Struct. (Theochem)*, **279**, 233 (1993).

[6] C. Adamo and V. Barone *in* "Chemistry and Properties of Biomolecular Systems", N. Russo, J. Anastassopoulou, and G. Barone (eds.), vol. II, 1 (1993), Kluwer Acad. Publ., Dordrecht, NL.

[7] O. Tapia, *Theor. Chim. Acta*, **47**, 157 (1978); O. Tapia and O. Goscinski, *Mol. Phys.*, **29**, 1653 (1975); M. W. Wong, K. B. Wiberg, and M. Frisch, *J. Chem. Phys.*, **95**, 8991, (1991).

[8] J. Tomasi, *Int. J. Quantum Chem.*, *QBS*, **18**, 73 (1991); S. Miertus, E. Scrocco, and J. Tomasi, *Chem Phys.*, **55**, 117 (1981); F. Floris and J. Tomasi, *J. Comp. Chem.*, **10**, 616 (1989).

Chapter 5

Electronic Charge Distributions

5.1 Introduction

One of the most popular concepts in chemistry is that of charge distributions in molecules. Pictorial presentations of charge densities can be offered in a number of ways. In the familiar contour map type [1], for example, contours corresponding to various values of the charge density (or of its difference with respect to the superposition of the free atoms) are plotted for different points in a specified plane of the molecule [2]–[6]. This type of presentation is certainly a realistic one, namely with electron densities calculated from Hartree–Fock wave functions which are generally known to be reasonably accurate. Contour maps alone, however, do not tell us how much charge can be assigned in a meaningful way to the individual atoms of a molecule. Hence our problem: the picking of the right atomic charges.

It is by no means an easy task to search for the correct answer[1]. Many attempts have been made [6] but, for some reason or other, the numbers do not seem to come out right, certainly not all of them, assuming that one knows what 'right' means in this context. Briefly, it looks as if the problem of apportioning the electronic charge of a molecule among its atoms was going to stump the theory of bond energies.

As long as atomic charges are taken as by-products of molecular calculations and used as an interpretative adjuvant in semi-quantitative discussions of chemical problems, the lack of well-justified numerical results may be per-

---

[1] Confucius, and others.

ceived, say, as an annoyance but does not represent an acute problem in itself. Here things are different. So we pick up the numbers from scratch and ask ourselves what 'some reason or other' means in the present context. Pursuing this idea with the help of results obtained from conventional methods, it becomes possible to get some useful answers, although the general solution for theoretical, physically justified charge analyses still remains in the outskirts, unfortunately. Yet, the fragmentary insight gained in this way does permit significant applications and verifications of the theory and should hopefully encourage future research in this area.

The conventional starting-point is well known: atomic charges are most conveniently calculated from molecular wave functions following Mulliken's population analysis [7]. Results derived therefrom are a solid basis for the discussion of important aspects regarding static charge distributions in the Born–Oppenheimer approximation, in spite of the fact that one should not put too much reliance on numbers calculated in this way without a detailed consideration of their basis set dependence. Appropriate modifications are considered at a later stage.

5.2   Mulliken's population analysis

The knowledge of the molecular wave function enables us to determine the electron density at any given point in space. Here we inquire more specifically about the amount of electronic charge that can be associated in a meaningful way with each individual atom of an N-electron system. Mulliken's population analysis [7] for extracting atomic charges from wave functions is rooted in the LCAO formulation and is not directly applicable to other types of wave functions. If the ith normalized molecular orbital is written

$$\phi_i = c_{Ai}\phi_A + c_{Bi}\phi_B + \cdots + c_{Mi}\phi_M$$

as a linear combination of normalized atomic orbitals centered on atoms A, B, ... with real coefficients, the probability density associated with an electron in this MO is

$$|\phi_i|^2 = c_{Ai}^2\phi_A^2 + c_{Bi}^2\phi_B^2 + \cdots + 2c_{Ai}c_{Bi}\phi_A\phi_B + 2c_{Ai}c_{Ci}\phi_A\phi_C + \cdots$$

Integrating this equation over three-dimentional space and using the fact that ϕ_i and the ϕ_A, ϕ_B, ... atomic orbitals are normalized, we get

$$1 = c_{Ai}^2 + c_{Bi}^2 + \cdots + 2c_{Ai}c_{Bi}S_{AB} + 2c_{Ai}c_{Ci}S_{AC} + \cdots$$

where S_{AB}, $S_{AC}, \ldots$ are the overlap integrals defined (in conventional notation) as $S_{AB} = \langle \phi_A | \phi_B \rangle$, $S_{AC} = \langle \phi_A | \phi_C \rangle$, etc.

Mulliken's analysis associates $\nu_i c_{Ai}^2$ electrons with atom A, $\nu_i c_{Bi}^2$ with atom B, etc., in the ith MO, where ν_i is the occupation number of that MO. The overlap populations

$$2\nu_i c_{Ai} c_{Bi} S_{AB}, \quad 2\nu_i c_{Ai} c_{Ci} S_{AC}, \ldots$$

are divided equally between the centers A and B, A and C, $\ldots$, respectively. The population arising from the ith molecular orbital is thus, e.g., for A

$$\nu_i(c_{Ai}^2 + c_{Ai} c_{Bi} S_{AB} + c_{Ai} c_{Ci} S_{AC} + \cdots)$$

and the total number of electrons associated with atom A is obtained from a summation over all the occupied molecular orbitals, with similar expressions for the other atoms.

This analysis is readily generalized to molecular orbitals which have more than one component atomic orbital with a given atomic center. With $c_{r_k i}$ representing now the coefficient of the rth type of atomic orbital ($1s, 2s$, etc.) of atom k in the ith molecular orbital, we describe the latter by

$$\phi_i = \sum_{r_k} c_{r_k i} \chi_{r_k}$$

where the summation extends over all the appropriate normalized basis functions χ_{r_k} and the subindex k labels the different nuclei in the system. The corresponding overlap population associated with atoms k and l due to atomic orbitals of type r and s, respectively, is then

$$2\nu_i c_{r_k i} c_{s_l i} S_{r_k s_l} \quad \text{with} \quad S_{r_k s_l} = \langle \chi_{r_k} | \chi_{s_l} \rangle.$$

Finally, Mulliken's analysis yields the population, N_k, on atom k from the appropriate sums over all doubly occupied molecular orbitals i and over all types of basis functions, that is

$$N_k = \sum_i \sum_r \nu_i \left(c_{r_k i}^2 + \sum_{l \neq k} c_{r_k i} c_{s_l i} S_{r_k s_l} \right). \tag{5.1}$$

The quantity which is referred to in most cases is the net atomic charge q_k defined as

$$q_k = Z_k - N_k \tag{5.2}$$

where Z_k is the nuclear charge of center k. With this definition, q_k is negative when an atom in a molecule carries a number of electrons in excess of that of the neutral atom. This definition is generally valid, independently of the mode of obtaining N_k. When N_k is derived from a Mulliken population analysis, as in (5.1), q_k is referred to as Mulliken net atomic charge.

5.3   Generalized population analysis

The N_k result expressed by eq. (5.1) implies the half-and-half partitioning of all overlap population terms among the centers k, l, ... involved. The division of the overlap charge has concerned many authors [8]–[10]. While the usual (Mulliken) half-and-half assignment is certainly defensible in situations involving partners of equal nature, it can be questioned in cases involving dissimilar atoms. Because of the large weight of the overlap contributions to the total atomic populations, even minor (say, $1-3\%$) departures from the usual half-and-half assignment would result in sizeable modifications of the calculated net atomic charges. This point must be borne in mind in any assessment of Mulliken-type electron distributions. Assuming now a modified (and more general) mode of distributing overlap populations, one obtains for the population on center k that

$$N_k = \sum_i \sum_r \nu_i \left(c_{r_k i}^2 + \sum_{l \neq k} c_{r_k i} c_{s_l i} S_{r_k s_l} \lambda_{r_k s_l} \right) \tag{5.3}$$

where the weighting factor $\lambda_{r_k s_l}$ causes the departure from the usual halving of the overlap terms. Mulliken's charges correspond to $\lambda_{r_k s_l} = 1$. One may employ the coefficients λ in order to conserve the dipole moment, for example, or (more generally) in order to define a scaling of atomic charges satisfying correlations with some property, such as ^{13}C nuclear magnetic resonance shifts, for example—an undertaking that amounts, in a way, to an experimental partitioning of overlap populations [9]–[11].

In terms of the difference

$$\sum_i \sum_r \sum_{l \neq k} \nu_i c_{r_k i} c_{s_l i} S_{r_k s_l} (1 - \lambda_{r_k s_l}) = \sum_{l \neq k} p_{kl} \tag{5.4}$$

between Mulliken charges, eq. (5.1), and those given by eq. (5.3), one obtains for the net atomic charge q_k of atom k, eq. (5.2), that

$$q_k = q_k^{\text{Mulliken}} + \sum_{l \neq k} p_{kl} \tag{5.5}$$

where the charge normalization $\sum_k N_k = N$ or, for a neutral molecule,

$$\sum_k q_k = 0 \tag{5.6}$$

is ensured by the condition $p_{lk} = -p_{kl}$.

Because of the sharply decreasing values of the overlap integrals with distance, the principal contributions by far to the sum $\sum_{l \neq k} p_{kl}$ are those due to the overlap terms involving the atom(s) l directly bonded to k. Moreover, it turns out that the overlap populations involving specified k, l atom pairs under comparable situations are practically the same in series of closely related compounds (e.g., the saturated hydrocarbons). *Under these conditions* we can approximate eq. (5.5) by assuming a constant p_{kl} correction for each type of k–l bond formed by atom k. For the hydrocarbons, which represent the class of molecules primarily aimed at, this simple approximation gives

$$q_C = q_C^{\text{Mulliken}} + N_{\text{CH}} \, p \tag{5.7}$$

$$q_H = q_H^{\text{Mulliken}} - p \tag{5.8}$$

with $\sum p_{\text{CH}} = N_{\text{CH}} p$, where p is the 'error' introduced by *one* H bonded to C and N_{CH} is the number of H atoms attached to that C. This first-order approximation gives us the tools for solving at least one problem, that of the charge distributions in hydrocarbons.

Equation (5.3) was presented here as a way of correcting shortcomings attributed to the half-and-half partitioning of overlap populations. Now we must remember that the selection of basis sets dramatically affects Mulliken charge results[2], suggesting that part, if not all, of (5.4) could be interpreted as a way of correcting for basis set effects. (This topic is developed in Section 5.) Mayer's analysis [16] vindicating Mulliken's halving of overlap terms strongly points in that direction and it is clear that the partitioning problem should not be discussed without explicit reference to the bases which are utilized in LCAO expansions.

Finally, one should also pay attention to configuration interaction and examine the role of singly and multiply excited configurations in Mulliken charge analyses.

---

[2]The charge of the methane carbon atom, for example, goes from -48.92 me (optimized STO–3G result [12]) to -790 me in $(7s3p|3s)$ calculations [13] and to -1072 me in Allen's calculation [14]. The point is that in variational energy calculations there is more to be gained by adding to the basis of a heavy atom than by an expansion of the hydrogen basis. Improved descriptions of H restore the situation, to the point that the carbon atom can be made positive [15].

5.4   Configuration interaction calculations

It is well known that the theoretical energies of ground-state molecules are but slightly improved by the inclusion of single excitations in CI calculations[3]. In contrast, the important role of single excitations is revealed [18] by the erroneous dipole moment predicted for CO when only multiple excitations are retained; inclusion of single excitations restores the correct sign and order of magnitude for μ_{CO}.

The sought-after improvement of the many-electron wave function is achieved by expanding it as indicated in eq. (1.48). Inclusion of the singly excited configurations is needed for a proper description of one-electron properties such as charge densities and, hence, for properties which are sensitive to this quantity (atomic charges, dipole, quadrupole moments,...). The doubly excited configurations are required for two reasons: i) they allow to indirectly introduce the single excitations by a coupling with the SCF reference state (the direct coupling being forbidden by the Brillouin theorem) and ii) they can lead to non-negligible contributions to charge density corrections through a direct coupling with the reference state.

Actual calculations [19] are rooted in the Rayleigh–Schrödinger perturbation theory, using a wave function expanded up to second order. The first-order correction $|\Psi_0^{(1)}\rangle$ allows only the doubly excited configurations to mix with the SCF reference state $|\Psi_0\rangle$, all other excitations making a zero contribution at this stage, while the second-order correction $|\Psi_0^{(2)}\rangle$ allows single excitations to mix indirectly with $|\Psi_0\rangle$ through the doubly excited configurations. In order to obtain the best possible wave function while keeping the required computer space at an acceptable level it is appropriate to select the configurations i) by retaining all singly excited symmetry- and spin-adapted configurations in the CI process and ii) by picking in order of preference those double excitations which mix more efficiently with single excitations and with the SCF reference state. The latter rule introduces a new constraint: one-electron properties are thus expected to converge more rapidly than the correlation energy. Hence a selection of configurations which may be appropriate in charge calculations does not necessarily represent the best choice in problems of correlation energies. Briefly, we approach convergence of the charge results well before achieving that of the energy.

---

[3]In the case of H_2O, for example, single excitations improve the correlation energy associated with double excitations (DCI) by only $\sim 0.6\%$ in a situation where single and double excitations (SDCI) account for $\sim 92\%$ of the basis set correlation energy (full CI) [17].

TABLE 5.1. Contributions of excited configurations to the correlation energy and the Mulliken net charge of carbon

Molecule	Level of calculation	ΔE_{corr}, au	Δq_C, me
Ethylene	SCF	0.	0.
	DCI	−0.10505	6.7
	SDCI	−0.10677	76.5
	Single contrib.	−0.00172	69.8
Acetylene	SCF	0.	0.
	DCI	−0.13468	2.0
	SDCI	−0.13678	64.1
	Single contrib.	−0.00209	62.1

These points are illustrated in Table 5.1 by the results obtained for ethylene (20 single + 283 double excitations) and acetylene (17 single + 340 double excitations), using an optimized 4–31G basis [19]. It is visible that the improvements of the energy, on the one hand, and charge distribution, on the other, follow distinct patterns. The modest contribution of the doubly excited configurations to the charges, in pure DCI calculations, shows that the main correction to the density primarily arises from interactions between single and double excitations and between single excitations themselves, which reflects in essence the one-electron nature of the density operator.

Selected charge results [19] are indicated in Table 5.2, for future use. It is clear that charge analyses, especially those of unsaturated systems, would be plagued by serious errors if CI corrections were left out.

TABLE 5.2. SCF and SDCI Mulliken (4–31G) carbon net charges, me

Molecule	Net charge on carbon	
	SCF	SDCI
Methane	−523.2	−515.7
Ethane	−382.7	−378.2
Ethylene	−346.4	−269.9
Acetylene	−335.3	−271.2

5.5   Charge analysis of simple alkanes

Let us begin with the chemist's ideas pertaining to changes of electron densities induced by substitution. They include the familiar order of electron-releasing ability of alkyl groups:

$$tert-C_4H_9 > \cdots > iso-C_3H_7 > C_2H_5 > CH_3.$$

This is the well-known inductive order. Correlations involving experimental (kinetic and equilibrium) data suggest a linear numerical ordering. In Taft's scale of polar σ^* constants (Table 5.3), the σ^* parameters are increasingly negative as the groups they describe are better electron donors [20, 21].

Our analysis considers an alkane as an alkyl group R attached to either CH_3 or H. Tentatively assuming the validity of the current interpretation of inductive effects described by Taft's scale, we write

$$q_{CH_3} = a\sigma_R^* \qquad (5.9)$$

$$q_H = a\sigma_R^* + b. \qquad (5.10)$$

Equation (5.9) is for alkanes described as $R-CH_3$ and q_{CH_3} is the net charge of a methyl group attached to R. Similarly, eq. (5.10) is for alkanes written as $R-H$ and q_H is the net charge of H. The validity of these equations is convincingly demonstrated by SCF Mulliken net charges. The example shown in Figure 5.1 uses fully optimized STO-3G charges [12]. This example is typical: *Mulliken charges given by every basis set satisfy (5.9) and (5.10)* [21]. Thus we could use the charges given by any LCAO basis of our choice to create an arbitrary set of scaling parameters (as a replacement for the σ^*'s) and proceed with (5.9) and (5.10) without invoking inductive effects. For convenience, however, we shall go along with the σ^*'s of Table 5.3.

TABLE 5.3. Polar σ^* constants

Group R	$\sigma^*(R)$	Group R	$\sigma^*(R)$
CH_3	0	neo-C_5H_{11}	-0.151
C_2H_5	-0.100	i-C_3H_7	-0.190
n-C_3H_7	-0.115	sec-C_4H_9	-0.210
n-C_4H_9	-0.124	$(C_2H_5)_2CH$	-0.225
i-C_4H_9	-0.129	$tert$-C_4H_9	-0.300

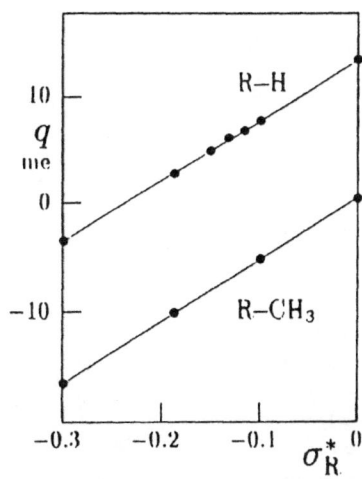

FIGURE 5.1. Verification of eqs. (5.9) and (5.10) by means of Mulliken net charges deduced from STO-3G calculations involving complete optimizations of geometry and orbital exponents. The net charge q is q_{CH_3} of (5.9) for the lower line and q_H of eq. (5.10) for the line R–H.

Charge results obtained from different bases are surely bewildering in their diversity. Equations (5.9) and (5.10) epitomize what these charges have in common and the reason they differ: the discussion of their basis set dependence now comes down to discussing the parameters a and b. This is where we can turn things to our advantage.

The idea is simple. Since SCF Mulliken charges satisfy (5.9) and (5.10), we can reverse the argument and deduce an internally consistent set of charges starting off with (5.9) and (5.10) and the σ^*'s of Table 5.3. Charge normalization

$$\sum_k q_k = 0 \qquad (5.11)$$

is part of this calculation. The general outline of this back-calculation of charges can be illustrated as follows, taking propane as an example. In this molecule there are four unknown charges, namely, those of the primary and secondary carbon atoms and of two different hydrogens (the weighted average of the primary H atoms is considered in this case). Because of (5.11), three equations are required for solving the problem of the three remaining unknowns. One of these equations is (5.9), taking propane as C_2H_5–CH_3, the other two are given by (5.10), considering propane as n-C_3H_7–H (for the calculation of the primary H atoms) or as i-C_3H_7–H (for the secondary H atoms). Hence we can solve the problem. Of course, this approach is by no means a general way of obtaining charges, but the fact that it can be applied to an adequate collection of molecules is sufficient for our intended purpose.

TABLE 5.4. Net charges in relative units

Molecule	Atom	Net charge
Methane	C	$4(n+1)/3n$
Ethane	C	1.000
Propane	C_{prim}	$(3n+0.55)/3n$
	C_{sec}	$(2n-3.8)/3n$
	H_{prim}	$(0.15-n)/3n$
	H_{sec}	$(0.9-n)/3n$
Butane	C_{prim}	$(3n+0.43)/3n$
	C_{sec}	$(2n-3.35)/3n$
	H_{prim}	$(0.24-n)/3n$
	H_{sec}	$(1.1-n)/3n$
Pentane	C_{centr}	$(2n-2.8)/3n$
	H_{centr}	$(1.25-n)/3n$
Isobutane	C_{prim}	$(3n+1.03)/3n$
	C_{tert}	$(n-7.7)/3n$
	H_{prim}	$(0.29-n)/3n$
	H_{tert}	$(2-n)/3n$
Neopentane	C_{prim}	$(n+0.49)/n$
	C_{quat}	$-4/n$
	H	$(0.51-n)/3n$

To begin with, let the carbon net charge of ethane $= 1$ arbitrary unit and express the other charges with respect to that reference. Then write

$$a = -10/3n \qquad (5.12)$$

because this expression for the slope a in terms of the new variable n is convenient in the presentation of the final results. Eq. (5.9) thus becomes

$$q_{CH_3} = -(10/3n)\sigma_R^*. \qquad (5.13)$$

On the other hand, eq. (5.10) shows that the ethane-H net charge, $-\frac{1}{3}q_C^{C_2H_6}$, is $-\frac{1}{3} = b - 0.1a$ (in arbitrary units); hence we get from (5.12) that

$$b = -(n+1)/3n. \qquad (5.14)$$

We know from (5.10) that b is the hydrogen net charge in methane because $\sigma_R^* = 0$ for R=CH_3. Using (5.12) and (5.14), we can rewrite (5.10) as follows

$$q_H = -(10\sigma_R^* + n + 1)/3n. \qquad (5.15)$$

Finally, eqs. (5.11), (5.13), (5.15) and the σ_R^*'s of Table 5.3 give the results listed in Table 5.4 [21] by reference to the unit charge defined by $q_C^{C_2H_6} = 1$.

TABLE 5.5. Optimized STO-3G Mulliken net
charges and charges deduced for $n = 1.3325$

Molecule	Atom	STO-3G	$n = 1.3325$
Methane	C	−48.92	−48.92
Ethane	C	−20.96	−20.96
Propane	C_{prim}	−23.81	−23.84
	C_{sec}	5.94	5.95
	H_{prim}	6.23	6.20
	H_{sec}	2.20	2.27
Isobutane	C_{prim}	−26.39	−26.36
	C_{tert}	33.36	33.39
	H_{prim}	5.50	5.47
	H_{tert}	− 3.53	− 3.50
Neopentane	C_{prim}	−28.66	−28.67
	C_{quat}	62.92	62.92

Table 5.4 is a compact presentation of charge results given by any LCAO-MO method. Tests are straightforward. The following example is worked out for Mulliken charges calculated from optimized STO-3G wave functions. Using the charges of methane and ethane, −48.92 and −20.96 me, respectively, it is $(-48.92)/(-20.96) = 4(n+1)/3n$, which gives $n = 1.3325$. We apply this n in Table 5.4 and obtain the charges in relative units, which must be multiplied by −20.96 to give the results in me units (Table 5.5). Tests of this sort are equally conclusive for semi-empirical (e.g., INDO [22], PCILO [23] and Extended Hückel [24]) as for large Gaussian basis-set calculations [9, 21][4]. What differs from method to method are n and the ethane-C net charge. A GTO($6s3p|3s$) basis, for example, leads to $n = 14.11$ and $q_C^{C_2H_6} = -232.5$ me [12], whereas Leroy's ($7s3p|3s$) calculations [13] correspond to $n = 42.3$ and $q_C^{C_2H_6} = -573$ me.

This brings up the obvious question: what are the 'true' values of n and $q_C^{C_2H_6}$? The generalized population analysis (Sect. 3) is part of the answer.

---

[4]The quality of the agreement depends on the precision of calculated SCF charges. Results are consistently improved by careful geometry optimizations and, most importantly, when all exponents, including those of the carbon K shells, are optimized individually for each molecule and each non-equivalent atom in the same molecule until stable charges (say, within ∼ 0.01 me) are obtained [9, 10, 21]. Practically, this is feasible only with small basis sets. On the other hand, the requirement for this sort of detailed ζ-optimizations appears to be somewhat less stringent with the use of extended basis sets.

Meaningful atomic charges

The meaning of n can be inferred from (5.12) where a measures, in a way, the sensitivity of charge variations to substituent effects. Small $|n|$ values indicate strong substituent effects. If inductive effects did not exist, the charges would be those corresponding to $|n| = \infty$, i.e., $a = 0$, and all H atoms would carry the same charge. No theoretical method leads to this extreme result.

The sign of a is of utmost importance. Equation (5.10) shows that a must be positive in order to reflect the usual order of electron-releasing abilities *tert*-$C_4H_9 > \cdots > CH_3$ because only then will a hydrogen atom attached to a *tert*-butyl group be electron-richer than that of methane. Similarly, as indicated by (5.9), only then will the methyl group in propane carry a net negative charge[5]. Eq. (5.12) gives a in arbitrary units. The corresponding expression in charge units is

$$a = - \left(\frac{10}{3n}\right) q_C^{C_2H_6} \qquad (5.16)$$

so that $q_C^{C_2H_6}$ and n must be of opposite signs to satisfy the constraint $a > 0$. *Ab initio* Mulliken charges usually[6] correspond to $n > 0$ and $q_C^{C_2H_6} < 0$, as is the case with most semi-empirical results, INDO being the notable exception with $n = -2$ and $q_C^{C_2H_6} = +71$ me [9]. It is worth remembering that *any LCAO-MO method leading to an n value with the same sign as the carbon net charge of ethane is bound to describe charge variations in the wrong order.* The test is easy: it suffices to calculate methane and ethane and get n.

We can now ask how the modified charges q_C and q_H of eqs. (5.7) and (5.8) compare with their original (Mulliken) counterparts. Let us define q_C^{oMull} = ethane-C Mulliken net charge, with $n = n^{Mull}$, and q_C^o = modified ethane-C net charge, corresponding to $n = n$. Equations (5.10) and (5.16) tell us that the slopes of the hydrogen net charges *vs.* σ^* are $-10q_C^o/3n$ for the modified charges and $-10q_C^{oMull}/3n^{Mull}$ for the original Mulliken charges. It is clear, however, that the transformation shown in (5.8) leaves the slope of the hydrogen net charge *vs.* σ^* unaffected, thus indicating that

$$\frac{q_C^o}{n} = \frac{q_C^{oMull}}{n^{Mull}}. \qquad (5.17)$$

---

[5]Pertinent experimental evidence for the $(CH_3)^- - (C_2H_5)^+$ polarity is offered in [25].
[6]Addition of diffuse functions to the basis may result in $n < 0$ and $q_C^{C_2H_6} > 0$ [15].

Finally, according to eq. (5.7), we have

$$p = \frac{1}{3}\left(q_C^\circ - q_C^{\circ\text{Mull}}\right) \qquad (5.18)$$

so that

$$p = \frac{n - n^{\text{Mull}}}{3n^{\text{Mull}}}\, q_C^{\circ\text{Mull}}. \qquad (5.19)$$

Equations (5.17)–(5.19) enable us to translate any set of atomic charges corresponding to well-defined values of n and $q_C^{C_2H_6}$ into an equivalent set corresponding to a different value of n. Usually, we start off with Mulliken charges, apply (5.19) and proceed with eqs. (5.7) and (5.8). Note that formulas like those of Table 5.4 are not required from here on, but we still have to figure out the correct value of n to get p. Charges calculated for a wrong n are of no use in applications to physical problems: they are disqualified from the outset because *the carbon charges corresponding to any arbitrarily selected* n *are not linearly related to those of the 'true'* n [21].

There is one particular n that merits attention. It reflects customary ideas: "the electron-attracting power of otherwise similar atoms decreases as their electron populations increase, thus opposing charge separation". This concept views local charge variations as events occurring 'most reluctantly', suggesting that the carbon atoms found in alkanes should be very similar to one another, conceivably differing as little as possible from one another. For a set of alkanes, each one containing two different carbons with net charges q_r and q_s, this constraint amounts to minimizing the sum $\sum(q_r - q_s)^2$ over the set. Using the formulas of Table 5.4, it is found that

$$\frac{d\sum(q_r - q_s)^2}{dn} = 0$$

is satisfied by $n \simeq -4.4$. Alternatively, using Mulliken charges (e.g., those of Table 5.5) one finds the p value that generates a set of charges as similar as possible to a constant, say, Q, from the condition

$$\sum\left(q_C^{\text{Mull}} + N_{\text{CH}}p - Q\right)^2 \qquad \text{minimum.}$$

The charges of Table 5.5 yield $p \simeq 30$ me and (5.19) gives $n \simeq -4.4$. Of course, this result for n does not depend on the particular set of charges used in this type of calculation: only p is basis-set dependent, not the n deduced from this minimization of charge variations.

TABLE 5.6. Mulliken net charges and final net charges, eq. (5.7), me

Calculation	Mulliken net charge of carbon			Final C net charge	
	Methane	Ethane	Ethylene	Ethane	Ethylene
STO-3G	−48.92	−20.96	−128.4	69.40	−68.2
STO-3G+CI	−46.37	−22.28		55.14	
4-31G	−523.2	−382.7	−346.4	42.8	−62.7
4-31G+CI	−515.7	−378.2	−269.9	37.8	7.5
Empirical				35.1	7.7

Evidently, one cannot prove *a priori* that the postulate requiring minimal charge variations should apply. *A posteriori*, it appears as the only overture to charges which qualify in applications to physical problems. Indeed, comparisons involving experimental results indicate that $n = -4.4122$ (from ^{13}C NMR shifts, Section 6), $n = -4.41$ (from atomization energies [26], Chapter 6.5) and $n = -4.4083$ (from ionization potentials [27]).

We select the value $n = -4.4122$ with a precision that is not warranted by our actual knowledge but used in order to avoid disturbing rounding-off errors. This result and our charge formulas combine to give[7]

$$p = \frac{3.4122}{3}q_C^{\text{Mull}}(C_2H_6) - \frac{4.4122}{4}q_C^{\text{Mull}}(CH_4). \qquad (5.20)$$

At last we can apply our modified population analysis.

Table 5.6 describes ethane and ethylene showing, namely, their final charges obtained from (5.7) and the p value given by (5.20). The quality of the results derived from SDCI calculations is self-explanatory: the 4-31G+CI charges are remarkably close to their empirical counterparts (discussed in Section 7) which are optimum values in energy calculations (Chapter 6).

The negative n value found in this analysis means that the alkane carbon charges are positive. This relatively important C^+-H^- polarity is in line with the view that hydrogen is certainly more electronegative than carbon, as Mulliken and Roothaan have pointed out [28]. Since then, the C^+-H^- polarity has been advocated on various occasions, e.g., by Julg [29], Bader [30], Jug [8], Wiberg [31], and in localized MO theory [32]. Here it is a consequence of keeping carbon charge variations at a minimum.

---

[7]Using the formula for the methane C atom given in Table 5.4 and eq. (5.7), we get $4(-3.4122)/3(-4.4122) = [q_C^{\text{Mull}}(CH_4) + 4p]/[q_C^{\text{Mull}}(C_2H_6) + 3p]$ and thus (5.20).

5.6 Charge analyses using NMR shifts

Our interest in correlations between nuclear magnetic resonance shifts and atomic electron populations originates in the quest of practical means for obtaining atomic charges.

This sort of approach postulates that one of the major factors governing the shielding of a specific nucleus is its local electron density. Now, chemical shift is a property of the interaction of the charge density with an external magnetic field. It depends therefore on the value of the integrated charge density (or 'charge') in the neighborhood of a nucleus—as well as on other factors, of course, e.g., the magnetic susceptibility of that charge density— but a formal relationship between NMR chemical shifts and atomic charges is not part of the rigorous theory of nuclear magnetic resonance.

The plausibility of charge–shift correlations is revealed by detailed analyses of the average diamagnetic and paramagnetic contributions, σ^d and σ^p, respectively, to the total average magnetic shielding $\sigma = \sigma^d + \sigma^p$. In series of closely related molecules, it turns out that any change in total (dia- plus para-) magnetic shielding of a heavy atomic nucleus (C, N, etc.) is nearly that of the *local* paramagnetic term, because of important cancellation effects involving nonlocal dia- and paramagnetic contributions [33, 34], i.e.,

$$\Delta\sigma_{\text{total}} \simeq \Delta\sigma^P_{\text{local}} \qquad (5.21)$$

where the impact of residual nonlocal terms is minimized in comparisons between atoms of the same type. Thus it makes sense to inquire about correlations between NMR shifts and atomic charges which, of course, are strictly local properties; but the practical answer stems ultimately from the actual examination of shift *vs.* charge results. As for now, we are restricted to using SCF results to pursue this objective.

Chemical shifts, δ, of ^{13}C, ^{17}O, etc., nuclei are usually given in parts per million (ppm) with reference to tetramethylsilane (TMS). Positive δ values indicate downfield shifts. In writing a linear relationship

$$\delta = \alpha q + \delta^\circ \qquad (5.22)$$

one must keep in mind that q becomes more negative as the corresponding electron population increases, eq. (5.2). A negative slope α means that a gain of electronic charge on a center translates into a downfield shift, i.e., into a larger δ value. Both positive and negative slopes are met in applications of eq. (5.22). For our use, it is convenient to rewrite (5.22) as follows

$$\delta^{\text{ref}} = \alpha\Delta q \qquad (5.23)$$

where Δq and δ^{ref} measure how the net charge and the chemical shift of a given atom differ from the charge and shift of an appropriately selected reference atom. (For the alkanes, for example, chosing the ethane carbon atom as reference, we have $\Delta q_C = q_C - q_C^{C_2H_6}$ and $\delta^{\text{ref}} = \delta_C - \delta_C^{C_2H_6}$.)

In π-systems, we separate the variations of σ charges, Δq_σ, from those of π charges, Δq_π, and write

$$\delta^{\text{ref}} = \alpha_\sigma \Delta q_\sigma + \alpha_\pi \Delta q_\pi. \tag{5.24}$$

Of course, $\Delta q = \Delta q_\sigma + \Delta q_\pi$. A situation which is frequently encountered, e.g., with ethylenic and aromatic carbon atoms [33] and with the nitrogen atoms found in molecules like pyridine, 1,2-diazine, 1,3-diazine and 1,3,5-triazine [35], is that σ and π populations vary regularly in the reverse order,

$$\Delta q_\sigma \simeq m \Delta q_\pi \quad \text{with} \quad m < 0,$$

so that $\Delta q \simeq (m + 1)\Delta q_\pi$ and $\delta^{\text{ref}} \simeq (m\alpha_\sigma + \alpha_\pi)\Delta q_\pi$. The latter two expressions and eq. (5.23) combine to give

$$\alpha \simeq (m\alpha_\sigma + \alpha_\pi)/(m + 1). \tag{5.25}$$

For the slope of δ *vs.* π electronic charge variations, at last, one obtains:

$$d\delta/dq_\pi \simeq m\alpha_\sigma + \alpha_\pi. \tag{5.26}$$

The central question now revolves about α and its calculation. The main trends revealed by SCF population analyses can be summarized as follows [33]. Any gain in total charge dictated by that of σ populations (i.e., $\Delta q_\pi = 0$ or very small) translates into a downfield shift, $\alpha < 0$, which is the trend exhibited by sp^3-hybridized carbon, carbonyl carbon, and dialkylether oxygen atoms, as well as by the nitrogen atoms found in alkylamines, nitroalkanes and isonitriles. On the other hand, any increase in total population resulting from a gain in π charge prevailing over a concurrent loss of σ electrons, i.e., $-1 < m < 0$, is accompanied by an upfield shift, $\alpha > 0$, which is typical for ethylenic and aromatic sp^2 carbon atoms, carbonyl oxygen atoms and the nitrogen atoms of azines (e.g., substituted pyridines).

Relationships involving sp³ carbon atoms

A most instructive comparison using the STO-3G Mulliken net charges of alkanes (Table 5.5) is shown in Figure 5.2. The points for the CH_3-, CH_2- and $CH-$ carbon atoms, including those of cyclohexane [36] and adamantane [37], lie on parallel, equidistant lines shifted from one another by ~ 30 me.

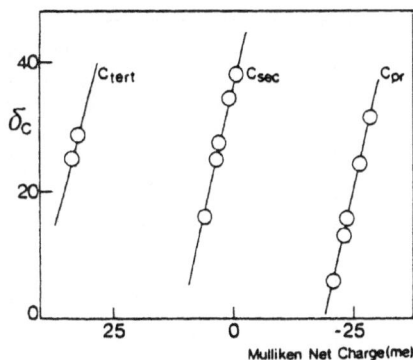

FIGURE 5.2. Comparison between the carbon-13 NMR shifts (ppm from TMS) of selected primary, secondary and tertiary sp^3 C atoms and Mulliken net charges from fully optimized STO-3G calculations [9].

Figure 5.2 suggests how the three lines can be made to merge into one single correlation line. Indeed, considering eq.(5.7), we write (5.22) as follows

$$\delta = \alpha(q_C^{Mulliken} + N_{CH}p) + \delta^\circ \qquad (5.27)$$

and deduce p from a multiple regression analysis using Mulliken charges. For the STO-3G charges used in this example, $p = 30.12$ me[8]. Figure 5.3 displays the results obtained from 'corrected' charges, eq. (5.7). This correlation is described by

$$\delta = -237.1(q_C/q_C^\circ) + 242.64 \text{ ppm from TMS}$$

with a standard error of 0.3 ppm. Using ethane as reference, (5.23) becomes

$$\delta^{ref} = -237.1(\Delta q_C/q_C^\circ) \text{ ppm from ethane}$$

and, with $q_C^\circ = 35.1$ me (Table 5.6),

$$\Delta q_C = -0.148\,\delta^{ref} \text{ me}. \qquad (5.28)$$

This is probably the most accurate charge–shift correlation which is presently known. Its validity was carefully examined for paraffins, cyclohexane and for molecules consisting of several chair or boat cyclohexane rings [37]. No extra effect appears to contribute to the chemical shift specifically because of the occurrence of six-membered cyclic structures. Similar conclusions do not apply to smaller cycles (e.g., cyclopropane) which fail to obey eq.(5.28). Changes in local geometry affecting the hybridization of carbon are certainly part of the problem [9], but no final conclusion can yet be drawn for these systems without additional information.

---

[8] Alternatively, one can reformulate the problem using the charge functions given in Table 5.4 and calculate n. The result, $n = -4.4122$, used in (5.19), leads to $p = 30.12$ me. Remember that p is basis set dependent, but n is not.

FIGURE 5.3. Correlation between ^{13}C NMR shifts (ppm from TMS) of sp^3 carbon atoms and atomic charges derived from eq. (5.7) using fully optimized STO-3G charges with $p = 30.12$ me. The charges are expressed in terms of q_C/q_C° and correspond to $n = -4.4122$. This figure includes all the points represented in Figure 5.2 plus methane and the quaternary C atom of neopentane. Note that $q_C/q_C^\circ < 1$ indicates a decrease of positive charge, i.e., a gain of electronic charge.

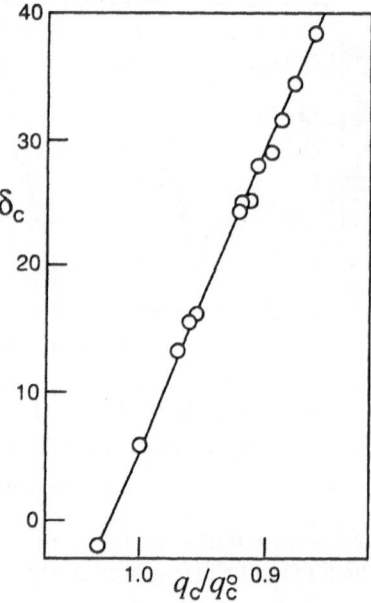

Relationships involving sp^2 carbon atoms

Selected results are presented in Table 5.7 for typical olefinic carbon atoms. The Mulliken net charges [11] follow from optimized STO-3G computations.

TABLE 5.7. Atomic charges, me, and NMR shifts of olefinic carbon atoms

Molecule	Atom	$q_{C,\sigma}^{Mull.}$	$q_{C,\pi}^{Mull.}$	$q_C^{Mull.}$	$q_C^{eq.(5.7)}$	δ
Ethylene	C (4)	−128.4	0.0	−128.4	−68.2	122.8
Propene	C–1 (2)	−120.4	−34.2	−154.6	−94.4	115.0
	C–2 (9)	−83.5	25.2	−58.3	−28.2	133.1
Isobutene	C–1 (1)	−112.2	−61.7	−173.9	−113.7	109.8
	C–2 (10)	−55.2	47.6	−7.6	−7.6	141.2
trans-Butene	C–2 (7)	−78.6	−7.7	−86.3	−56.2	125.8
cis-Butene	C–2 (6)	−77.6	−8.8	−86.4	−56.3	124.3
2-Methyl-2-butene	C–2 (8)	−49.8	12.3	−37.5	−37.5	131.4
	C–3 (3)	−73.8	−32.6	−106.4	−76.3	118.7
2,3-Dimethyl-2-butene	C–2 (5)	−50.2	−11.0	−61.2	−61.2	123.2

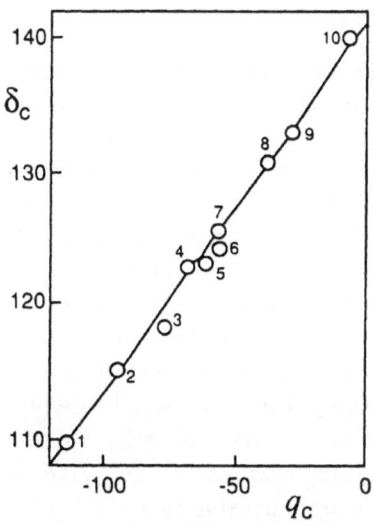

FIGURE 5.4. Comparison of ^{13}C NMR shifts, ppm from TMS, with the corrected carbon total $(\sigma+\pi)$ net charges, eq.(5.7), reported in Table 5.7 for selected olefins. The atom numbering is that indicated in Table 5.7. The radius of the circles correspond to an uncertainty of 0.7 ppm or 3.5 me. (Reproduced from [9].)

The comparison between the total net atomic charges, $q_C^{Mull.}$, and the experimental shifts [38] of olefinic sp^2 carbon atoms yields a result similar to that shown in Figure 5.2, but with $a > 0$. Again, we have three equidistant parallel regression lines, depending on the number of H atoms attached to the olefinic carbon. The analysis by means of eq.(5.27), on the other hand, leads to $p = 31.8 \pm 4$ me, that is, in essence, the value found for the sp^3 carbons, 30.12 me. Using the latter in eq.(5.7), we obtain the correlation displayed in Figure 5.4. The correlation between NMR shifts and SCF charges given by (5.7) is convincing. Still, SCF charges like those of Table 5.7 are unfit for realistic direct evaluations of the parameter α occurring in (5.22): CI corrections are a must in this sort of applications. Assuming that a relationship like (5.22) holds for the 'true' charges as well, a crude indirect estimate of α can be obtained from the values $m \simeq -0.955$ and $(d\delta_C/dq_\pi) \simeq 300$ ppm/e, giving $\Delta q_C \simeq 0.15\delta^{ref}$, which seem to be consistent with energy calculations (Chapter 6.6).

Good-quality charge–shift correlations are also known for aromatic [39] and carbonyl carbon atoms [9, 40]. For the aromatics—as with the olefinic C atoms, and for the same reasons—direct numerical applications of (5.22) are difficult: a tentative estimate of α from (5.25) and (5.26) is presented in Chapter 6.6. For the carbonyl carbon atoms—whose charges, like those of the alkanes, vary only at the $2s$ level—we can tentatively use (5.28) [9, 40].

Relationships involving oxygen atoms

The net atomic charges of the dialkylether-oxygen atoms mirror the progressive build-up of electronic charge with increasing electron-releasing ability of the substituents, in nice accord with the customary inductive effects of the alkyl groups: the oxygen atom in, say, di-*tert*-butylether is electron richer than that of dimethylether. Any gain in electronic charge at the oxygen atom is accompanied by a downfield ^{17}O NMR shift, satisfying eq. (5.22) with $\alpha < 0$ (Figure 5.5). So far, the general charge–shift trend characterizing these oxygen atoms are those described earlier for the sp^3 carbon atoms, with a difference, however. The orbital populations of the ether oxygen atoms [40] reveal that the population changes occur at the $2p$ level, not at the $2s$ level as in the alkane carbon atoms. Numerical analyses indicate [40] that the δ_O vs. q_O slope is $\sim 1/1.8$ that of the corresponding δ_C vs. q_C slope, obtained from the same basis set, i.e., $\Delta\delta_O/\Delta q_O \simeq \frac{1}{1.8}(\Delta\delta_C/\Delta q_C)$. Therefore, using now (5.28), we get the following approximation

$$\Delta q_O \simeq -0.267\,\delta_O \quad \text{me} \tag{5.29}$$

for the ether oxygen atoms. This approximation proves satisfactory in energy calculations (Chapter 6.7). Similar analyses are difficult for the carbonyl-oxygen atoms [9, 40]. As indicated in Chapter 6.7, $\Delta q_O^{\text{carbonyl}} \simeq 2.7\,\delta_O$ me possibly represents an acceptable solution, with NMR shifts taken from [42].

FIGURE 5.5. Correlation between ^{17}O NMR shifts [41] (ppm downfield from water) and STO-3G Mulliken net charges (me) [40] of dialkylethers R_1OR_2, where $R_1{=}R_2{=}CH_3$ (**1**); $R_1{=}CH_3$, $R_2{=}C_2H_5$ (**2**); $R_1{=}CH_3$, $R_2{=}iso$-C_3H_7 (**3**); $R_1{=}CH_3$, $R_2{=}tert$-C_4H_9 (**4**); $R_1{=}R_2{=}C_2H_5$ (**5**); $R_1{=}C_2H_5$, $R_2{=}iso$-C_3H_7 (**6**); $R_1{=}C_2H_5$, $R_2{=}$ $tert$-C_4H_9 (**7**); $R_1{=}R_2{=}iso$-C_3H_7 (**8**); $R_1{=}iso$-C_3H_7, $R_2{=}tert$-C_4H_9 (**9**); $R_1{=}R_2{=}tert$-C_4H_9 (**10**). (Reproduced from [9, 40].)

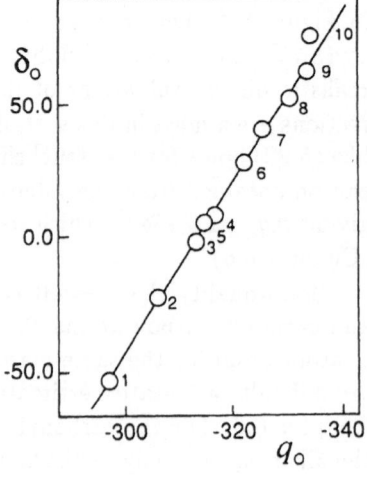

5.7 Selected reference atomic charges

Occasionally, charge normalization proves particularly helpful in the course of charge analyses. The idea is to calculate what can be calculated—usually, the alkyl part of a molecule—and to get the rest by difference. The value selected for the carbon atom of ethane, $q_C^{C_2H_6} = 35.1$ me, is the prime reference charge. It follows from energy calculations (Chapter 6.5), is reasonably well established and well supported by direct charge analyses (Table 5.6). The following observation is instrumental in this type of approach. In alkanes, the H and C charges are monotonically related to one another, with different lines for primary and secondary atoms. These lines are easily deduced using the charges of Table 5.4 with $q_C^{C_2H_6} = 35.1$ me, giving (in me units) 35.1 (C), −11.7 (H) for ethane; 33.64 (C), −12.10 (H) for propane-CH$_3$; 32.37 (C), −12.47 (H) for isobutane-CH$_3$, and 31.20 (C), −13.05 (H) for neopentane-CH$_3$. For the charges of CH$_2$ groups, the values are: 33.48 (C), −14.09 (H) for propane; 32.28 (C), −14.62 for butane, and 30.82 (C), −15.01 (H) for the central CH$_2$ of n-pentane.

In this vein, let us examine oxygen-containing molecules. (Carbon atoms adjacent to oxygen carry the label α.) Consider di-*tert*-butylether. For its CH$_3$ carbon atoms ($\delta_C = 26$ ppm from ethane), we have $q_C = 31.25$ and get $q_H = -13.02$ me, on the average, for its H atoms. On the other hand, using $\delta_{C_\alpha} = 7.7$ ppm from the α-carbon of diethylether, we get $q_{C_\alpha} = q_{C_\alpha}^\circ - 1.14$ and (using $\delta_O = 69.5$ ppm from diethylether [41]) $q_O = q_O^\circ - 18.56$ me, where $q_{C_\alpha}^\circ$ and q_O° are reference charges defined for diethylether. Thus we get from charge normalization that

$$q_O^\circ + 2q_{C_\alpha}^\circ \simeq 67.7 \quad \text{me}$$

for the ether oxygen and α-carbon atoms. Similarly, one obtains for acetone ($\delta_C = 22.2$ ppm from ethane for the methyl carbons [42]) that $q_C = 31.81$, $q_H = -9.40$ me and

$$q_{C_\alpha}^\circ + q_O^\circ \simeq -7.2 \quad \text{me}.$$

The problem now rests with the charges of the ether and acetone α-carbon atoms, $q_{C_\alpha}^\circ$. This is best approached using the difference

$$\Delta q_{C_\alpha}^\circ = q_{C_\alpha}^\circ - q_C^{\circ C_2H_6}$$

considering, however, that oxygen introduces an "extra" downfield shift at α-carbon (estimated [41] at ~ 41.7 ppm) which does not represent a carbon-charge effect. The latter is tentatively evaluated by subtracting 41.7 ppm from the observed α-carbon shifts (60, *viz.* 199 ppm from ethane for diethylether [41] and acetone [42], respectively) giving, with the help of (5.28), $\Delta q^o_{C_\alpha} \simeq -2.7$ me (ether) and -23 me (acetone). These estimates are admittedly very crude. A refinement, based on energy calculations [43], yields $\Delta q^o_{C_\alpha} \simeq -3.84$ me for diethylether and $\Delta q^o_{C_\alpha} \simeq -21.1$ me for acetone. Using these $\Delta q^o_{C_\alpha}$ values, it is found that $q^o_{C_\alpha} \simeq 31.26$ and $q^o_O \simeq 5.18$ me for diethylether, and $q^o_{C_\alpha} \simeq 14.0$ and $q^o_O \simeq -21.2$ me for acetone. The charges of the H atoms attached to α-carbons are deduced from charge normalization.

Now we turn attention to ethylene. Its carbon net charge is best evaluated using tetramethylethylene as a starter. The ^{13}C shift of its CH_3 carbons is 14.3 ppm from ethane, giving, with the help of (5.28), $\Delta q_C = -2.12$ me relative to the ethane carbon atom, i.e., $q_C = 35.1 - 2.12 = 32.98$ me. From the correlation between q_H and q_C net charges in methyl groups, q_H is estimated at -12.29 me. The net charge of the methyl group is therefore -3.89 me and that of the sp^2 carbon is 7.77 me. Tetramethylethylene (δ_C 123.2 ppm from TMS) and ethylene (δ_C 122.8 ppm from TMS) differ only by ~ 0.4 ppm, so that, assuming the validity of the approximation $\Delta q_C \simeq 0.15\,\delta^{ref}$, the ethylene carbon atoms are only marginally electron richer than the sp^2 carbons of tetramethylethylene. The final estimate for ethylene,

$$q_C^{ethylene} \simeq 7.7 \ \text{me},$$

is nicely supported by direct charge calculations (Table 5.6), as well as by a host of accurate energy calculations (Chapter 6.6).

In closing, let us examine benzene. The problem of calculating its charges is best approached by considering toluene. The ^{13}C NMR shift of its methyl carbon is 21.3 ppm from TMS (i.e., 15.7 ppm from C_2H_6). From eq. (5.28) we get $\Delta q_C \simeq -2.32$ me, so that $q_C \simeq 32.78$ me for this carbon atom. Using the correlation between methyl-C and methyl-H net charges, the latter are estimated at ~ -12.32 me. The toluene CH_3 group is, in this approximation, ~ -4.18 me negative. On the other hand, eqs. (5.9) and (5.10)[9] tell us that a H atom replacing a CH_3 group (under otherwise identical conditions) is ~ 9.05 me more negative than the latter. In this estimate, we find that the hydrogen atoms of benzene should be negative by ~ 13.2 me and, hence, that

$$q_C^{benzene} \simeq 13.2 \ \text{me}.$$

The at least approximate validity of this rough estimate is confirmed by applications in numerous energy calculations (Chapter 6.6).

---

[9]Remember that b stands for the hydrogen net charge of methane. Use (5.14) with $n = -4.4122$, then multiply the result by 35.1 to get b in me units.

5.8   Assessment

This considerably abridged account does not render justice to the great many studies which have dealt with the old problem of charge partitioning in polyatomic systems. One thing is clear, however: it is an involved, not-so-easy task to come up with realistic, physically meaningful answers concerning the distribution of electronic charge among the atoms forming a molecule.

Mulliken's well-known population analysis (Section 2) and its simple modification (Section 3) certainly represent the most straightforward approaches in calculations using LCAO wave functions. These approaches are rooted in solid theory and are the only ones considered in this discussion. They differ from one another in that the original Mulliken scheme consistently adopts the half-and-half partitioning of all overlap population terms among the centers involved, thus contrasting with the modified scheme that considers alternate ways of assigning overlap terms. Attention shall be given to criteria regarding the partitioning of overlap populations.

It is well established that numerical Mulliken charges strongly depend on the theoretical method (CNDO, INDO, Extended Hückel, etc.) used for their calculation. In *ab initio* SCF calculations, they strongly depend on the basis set employed. Here we consider only *ab initio* LCAO-MO results and thus treat numerical distortions of this sort as basis-set effects. These effects are large. And to top it all, one cannot ignore the role of single and multiple excitations in charge analyses, as revealed by calculations using CI wave functions (Section 4)—a circumstance that adds considerably to the difficulty of obtaining physically meaningful theoretical charges.

To help develop charge analyses, introduction of 'something new'—say, a suitable criterion for defining charge partitioning—is required at this point. The idea is to involve comparisons between calculated charges and some appropriate observable. This seems a natural thing to do if physical reality is attached to theoretical charges although, admittedly, in the absence of directly observable atomic charges unambiguously associated with the individual centers, we may experience difficulties in selecting (or justifying) what the 'appropriate observable' should be. NMR shifts are good candidates in this type of approach (Section 6) although the reasons for correlations between atomic charges and chemical shifts are rooted in intuition rather than in rigorous theory. Other observables called upon were molecular energies and ionization potentials. Extensive numerical verifications indicate that *all the examined properties*, NMR shifts, energies, etc., *definitely correspond to one and the same and surely unique definition of atomic charge.*

This result is satisfactory, of course, and follows from the general formula

$$q_k = q_k^{\text{Mulliken}} + \sum_{l \neq k} p_{kl}$$

assuming also that overlap populations involving specified k, l atom pairs in comparable situations are practically the same in series of closely related compounds. Little can be learned about p_{kl}—the departure from Mulliken's partitioning for one bonded atom pair kl—from comparisons involving atoms with constant (or nearly constant) $\sum_{l \neq k} p_{kl}$ sums, as for the ether-oxygen atoms. In contrast, with sums of p_{kl} terms assuming different values for different atoms, we can learn about p_{kl} using comparisons between charges and physical observables: in a way, we end up defining an empirical mode of partitioning overlap populations. The prime example is offered by saturated hydrocarbons. Using charges calculated for a given basis set, one gets only one answer for p_{kl}, whatever the property used for its determination. The numerical value of p_{kl} depends—as do the q_k^{Mulliken} charges—on the basis set, but the appropriately corrected q_k charges do not depend a geat deal on it. In this sense we may interpret the $\sum_{l \neq k} p_{kl}$ term (or at least a major part of it) as a correction for basis-set effects. The point is that one always gets precisely the same *relative ordering* for the net charges of saturated hydrocarbons—an ordering that is definitely basis-set independent (Section 5): it corresponds to minimal charge variations for the carbon atoms in this class of molecules. Only the *absolute* values of these charges are somewhat basis-set dependent, within acceptable limits. The absolute value for the ethane-C atom, 37.8 me, calculated from SDCI wave functions, is reasonably close to its empirical counterpart, 35.1 me, obtained from energy calculations and we need not worry about this sort of minor discrepancy.

While we have gained expert knowledge about charges in saturated hydrocarbons—those reported here are surely the only ones fit for use in energy formulas featuring net charges—the general problem of charge partitioning is still far from being solved. We know about the ethylene-C atom, $q_C = 7.5$ me from CI analyses using the generalized charge partitioning method, *vs.* 7.7 me from energy calculations, but are presently on thin ice with charges of the other sp^2 carbon atoms, especially as regards their charge-NMR shift correlations. (Fortunately, it turns out that the part that is poorly known contributes relatively little in energy calculations.) Off the beaten path, charge analyses based on simple normalization constraints, though lacking in elegance, can prove most efficient, at times, as indicated in Section 7: they nicely agree with results deduced from energy calculations.

Bibliography

[1] J. R. Van Wazer and I. Absar, "Electron Densities in Molecules and Molecular Orbitals", Academic Press, New York, 1975.

[2] M. Roux, M. Cornille, and L. Burnelle, *J. Chem. Phys.*, **37**, 933 (1972); M. Roux, *J. Chim. Phys.*, **57**, 53 (1960); *ibid.*, **55**, 754 (1958); M. Roux, S. Besnainou, and R. Daudel, *J. Chim. Phys.*, **53**, 218, 939 (1956).

[3] R. F. W. Bader, *in* "International Review of Science", Physical Chemistry, Series Two, vol. 1, A. O. Buckingham (ed.), Butterworths, London, 1975, pp 43–78; P. E. Cade, R. F. W. Bader, W. H. Hennecker, and I. Keaveny, *J. Chem. Phys.*, **50**, 5313 (1969); R. F. W. Bader and A. D. Bandrauk, *J. Chem. Phys.*, **49**, 1653 (1968); R. F. W. Bader, W. H. Henneker, and P. E. Cade, *J. Chem. Phys.*, **46**, 3341 (1967); R. F. W. Bader and W. H. Henneker, *J. Am. Chem. Soc.*, **88**, 280 (1966); *ibid.*, **87**, 3063 (1965).

[4] P. Coppens, *in* "International Review of Science", Physical Chemistry, Series Two, vol. 2, J. M. Robertson (ed.), Butterworths, London, 1975, pp 21–56; P. Coppens and E. D. Stevens, *Adv. Quantum Chem.*, **10**, 1 (1977); B. Dawson, "Studies of Atomic Charge Density by x-Ray and Neutron Diffraction. A Perspective", Pergamon Press, London, 1975; F. L. Hirschfeld and S. Rzotkiewicz, *Mol. Phys.*, **27**, 319 (1974); B. J. Ransil and J. J. Sinai, *J. Am Chem. Soc.*, **94**, 7268 (1972); B. J. Ransil and J. J. Sinai, *J. Chem. Phys.*, **46**, 4050 (1967).

[5] E. A. Laws and W. N. Lipscomb, *Israel J. Chem.*, **10**, 77 (1972).

[6] V. H. Smith, Jr., *Phys. Scripta*, **15**, 147 (1977); A. Julg, *Top. Curr. Chem.*, **58**, 1 (1975).

[7] R. S. Mulliken, *J. Chem. Phys.*, **23**, 1833, 1841, 2338, 2343 (1955).

[8] P.-O. Löwdin,*J. Chem. Phys.*, **21**, 374 (1953); *ibid.*, **18**, 365 (1950); E.
     R. Davidson, *J. Chem. Phys.*, **46**, 3320 (1967); K. Jug, *Theor. Chim.
     Acta*, **39**, 301 (1975); *ibid.*, **31**, 63 (1973).

[9] S. Fliszár, "Charge Distributions and Chemical Effects", Springer-
     Verlag, New York, 1983.

[10] S. Fliszár, *Can. J. Chem.*, **54**, 2839 (1976); S. Fliszár, A. Goursot, and
     H. Dugas, *J. Am. Chem. Soc.*, **96**, 4358 (1974).

[11] H. Henry and S. Fliszár, *J. Am. Chem. Soc.*, **100**, 3312 (1978).

[12] G. Kean and S. Fliszár, *Can. J. Chem.*, **52**, 2772 (1974).

[13] J. M. André, P. Degand, and G. Leroy, *Bull. Soc. Chim. Belg.*, **80**, 585
     (1971).

[14] J. E. Williams, V. Buss, and L. C. Allen, *J. Am. Chem. Soc.*, **93**, 6867
     (1971).

[15] V. Barone and S. Fliszár, to be published.

[16] I. Mayer, *Int. J. Quantum Chem.*, **23**, 341 (1983); I. Mayer, *Chem.
     Phys. Lett.*, **97**, 270 (1983).

[17] R. J. Bartlett and I. Shavitt, *Chem. Phys. Lett.*, **50**, 190 (1977) [Er-
     ratum: *ibid.*, **57**, 157 (1978)]; R. J. Bartlett and I. Shavitt, *Int. J.
     Quantum Chem. Symp.*, **11**, 165 (1977) [Erratum: *ibid., Symp.*, **12**,
     543 (1978)].

[18] F. Grimaldi, A Lecourt, and C. Moser, *Int. J. Quantum Chem.*, **51**, 153
     (1967).

[19] C. Mijoule, J.-M. Leclercq, M. Comeau, S. Fliszár, and M. Picard,
     Can. J. Chem., **70**, 68 (1992); S. Fliszár, J.-M. Leclercq, C. Mijoule,
     and S. Odiot, *in* "Applied Quantum Chemistry", V. H. Smith, Jr., H.
     F. Schaefer III, and K. Morokuma (eds.), p 395, D. Reidel Publishing
     Company, Dordrecht, NL, 1986; C. Mijoule, J.-M. Leclercq, S. Odiot,
     and S. Fliszár, *Can. J. Chem.*, **63**, 1741 (1985).

[20] R. W. Taft *in* "Steric Effects in Organic Chemistry", M. S. Newman
     (ed.), Wiley, New York, NY, 1956; R. W. Taft, *J. Am. Chem. Soc.*, **75**,
     4231 (1953).

[21] S. Fliszár, G. Kean, and R. Macaulay, *J. Am. Chem. Soc.*, **96**, 4353 (1974).

[22] D. R. Salahub and C. Sándorfy, *Theor. Chim. Acta*, **20**, 227 (1971).

[23] S. Diner, J.-P. Malrieu, and P. Claverie, *Theor. Chim. Acta*, **13**, 1 (1969); J.-P. Malrieu, P. Claverie, and S. Diner, *Theor. Chim. Acta*, **13**, 18 (1969); S. Diner, J.-P. Malrieu, F. Jordan, and M. Gilbert, *Theor. Chim. Acta*, **13**, 101 (1969); S. Fliszár and J. Sygusch, *Can. J. Chem.*, **51**, 991 (1973).

[24] R. Hoffmann, *J. Chem. Phys.*, **39**, 1397 (1963); J. M. Sichel and M. A. Whitehead, *Theor. Chim. Acta*, **5**, 35 (1966).

[25] V. W. Laurie and J. S. Muenter, *J. Am. Chem. Soc.*, **88**, 2883 (1966).

[26] S. Fliszár, *J. Am. Chem. Soc.*, **102**, 6946 (1980).

[27] H. Henry and S. Fliszár, *Can. J. Chem.*, **52**, 3799 (1974).

[28] R. S. Mulliken and C. C. Roothaan, *Chem. Rev.*, **41**, 219 (1947).

[29] A. Julg, *J. Chim. Phys.*, **53**, 548 (1956).

[30] R. F. W. Bader and H. J. T. Preston, *Theor. Chim. Acta*, **17**, 384 (1970).

[31] K. B. Wiberg and J. J. Wendoloski, *J. Comput. Chem.*, **2**, 53 (1981).

[32] M. S. Gordon and W. England, *J. Am. Chem. Soc.*, **94**, 5168 (1972); R. H. Pritchard and C. W. Kern, *J. Am. Chem. Soc.*, **91**, 1631 (1969).

[33] S. Fliszár, G. Cardinal, and M.-T. Béraldin, *J. Am. Chem. Soc.*, **104**, 5287 (1982).

[34] E. C. Vauthier, S. Fliszár, F. Tonnard, and S. Odiot, *Can. J. Chem.*, **61**, 1417 (1983); E. C. Vauthier, S. Odiot, and F. Tonnard, *Can. J. Chem.*, **60**, 957 (1982).

[35] M. Comeau, M.-T. Béraldin, E. Vauthier, and S. Fliszár, *Can. J. Chem.*, **63**, 3226 (1985).

[36] R. Roberge and S. Fliszár, *Can. J. Chem.*, **53**, 2400 (1975).

[37] G. Kean, D. Gravel, and S. Fliszár, *J. Am. Chem. Soc.*, **98**, 4749 (1976).

[38] A. J. Jones and D. M. Grant, reported as personal commumication in J. B. Stothers, "Carbon-13 NMR Spectroscopy", Academic Press, New York, 1972; G. B. Savitski, P. D. Ellis, K. Namikava, and G. Maciel, *J. Chem. Phys.*, **49**, 2395 (1968) (for propene); J. W. de Haan and L. J. M. van de Ven, *Org. Mag. Res.*, **5**, 147 (1973) (for *trans-* and *cis*-butene); R. A. Friedel and H. L. Retcofsky, *J. Am. Chem. Soc.*, **85**, 1300 (1963) (for 2-methyl-2-butene).

[39] S. Fliszár, G. Cardinal, and N. A. Baykara, *Can. J. Chem.*, **64**, 404 (1986).

[40] M.-T. Béraldin, E. C. Vauthier, and S. Fliszár, *Can. J. Chem.*, **60**, 106 (1982).

[41] C. Delseth and J.-P. Kintzinger, *Helv. Chim. Acta*, **61**, 1327 (1978).

[42] C. Delseth and J.-P. Kintzinger, *Helv. Chim. Acta*, **59**, 466 (1976); **59**, 1411 (1976).

[43] S. Fliszár and M.-T. Béraldin, *Can. J. Chem.*, **60**, 792 (1982).

Chapter 6

Applications

6.1 Introduction

This Chapter presents numerical applications. Atomization energies are calculated and compared with experimental results. Saturated, olefinic and benzenoid hydrocarbons are examined in detail, as well as selected ethers and carbonyl compounds.

These calculations require an evaluation of the $\sum_k \sum_l a_{kl} \Delta q_k$ part of the general energy formula (3.60). This evaluation involves charges. The a_{kl} parameters, eq. (3.35), are fairly large, usually between -0.2 and -0.6 kcal mol^{-1} me^{-1}. It it thus clear that the precision of the $\sum_k \sum_l a_{kl} \Delta q_k$ term heavily depends on the quality of the charges that are used: they play a decisive role. Those described for the alkanes leave no room for excuses if we fail to come up with the correct answer: in principle, i.e., if they are as good as claimed, the $\sum_k \sum_l a_{kl} \Delta q_k$ part of (3.60) should match experimental accuracy. This part of the theory is best examined with the help of saturated hydrocarbons. The small nonbonded contributions, E_{nb}, calculated from (3.11), are part of this verification. A clear assessment regarding the calculation of $\sum_k \sum_l a_{kl} \Delta q_k$, and its precision, is particularly important for this series of molecules because most organic molecules contain alkyl groups and because their contribution usually represents the preponderant part of $\sum_k \sum_l a_{kl} \Delta q_k$.

Now we come to the $\sum_{k<l} \varepsilon_{kl}^{\circ}$ part of eq. (3.60). Only two constants are required for the alkanes, namely ε_{CC}° for the carbon–carbon bonds and ε_{CH}° for the carbon–hydrogen bonds. These reference bond energies are given by (3.18). Their accurate evaluation is difficult but well-worth doing in order to

avoid easily recognizable systematic errors depending only on the numbers of CC and CH bonds in the molecule. In principle, i.e., if ε_{CC}° and ε_{CH}° are properly determined, the burden of the verification of (3.60) for the alkane series rests entirely with the evaluation of $\sum_k \sum_l a_{kl} \Delta q_k$. Yet there is an other compelling and far-reaching reason prompting us to look for accurate estimates of ε_{CC}° and ε_{CH}°, selected by reference to ethane. New types of bonds occur in unsaturated hydrocarbons, namely the CH bonds of ethylene and benzene and a host of new carbon–carbon single bonds, like those found in propene, 1,3-butadiene, toluene, styrene, diphenyl, etc. All these new CH and CC bonds are deduced from the ethane reference values, ε_{CH}° and ε_{CC}°, respectively. This is done by means of eq. (3.39), which takes care of the charge effects, and of eq. (3.55) which accounts for geometry changes and possible displacements of the centroids of π orbitals. It is thus important to start off with good ε_{CH}° and ε_{CC}° input values in order to fairly monitor the validity of our transformations, namely as regards the use of eq. (3.55).

Briefly, the alkanes are best suited for testing the $\sum_k \sum_l a_{kl} \Delta q_k$ part of (3.60). The unsaturated hydrocarbons, on the other hand, offer a most instructive insight into the performance of eq. (3.55), namely as regards the derivation of new reference bond energies satisfying eq. (3.37). Finally, the oxygen containing molecules offer examples of an other kind: they indicate how to proceed when the second-order contributions, $\frac{1}{2}(\partial^2 E^{vs}/\partial N^2)^{\circ} \Delta q$, appearing in a_{kl} cannot be neglected. The examples are worked out in detail and verification of the results is made easy.

All the calculations are checked against experimental atomization energies. Appendix A indicates how the latter are retrieved from thermochemical and spectroscopic information.

6.2   Basic theoretical parameters

This Section describes the evaluation of the basic parameters γ_k^{at}, γ_k^{mol}, γ^v, $(\partial E_k^{vs}/\partial N_k)^{\circ}$ and $(\partial^2 E_k^{vs}/\partial N_k^2)^{\circ}$, ready for use in our energy formulas.

The γ parameters

The γ parameters of the isolated atoms, γ_k^{at}, are readily obtained from eqs. (1.35) or (1.36). Selected SCF results[1] are indicated in Table 6.1. For hydrogen, of course, $\gamma = 2$ because of the virial theorem.

---

[1]The atomic wave functions are from [1]. Additional results are reported in [2, 3].

TABLE 6.1. Selected SCF
γ_k^{at} and γ_k^{mol} parameters

Atom	γ_k^{at}	γ_k^{mol}
H	2	2
B	2.31965	2.31428
C	2.33863	2.33263
N	2.35905	2.34742
O	2.38039	2.37361
F	2.40096	2.39866

For atoms in a molecule, we use eq. (2.15) and rewrite it as follows

$$E_k = \frac{1}{\gamma_k^{mol}} \left(V_{ne,k} + Z_k \sum_{l \neq k} \frac{Z_l}{R_{kl}} \right) \tag{6.1}$$

with the help of (2.5) and (2.6). These γ's are obtained [2–6] from SCF potentials at the individual nuclei k, l, ... and from their fit with total energies using (2.17) and (6.1), assuming a constant γ_k^{mol} for each atomic species. The latter point merits attention because, as seen in (6.1), $1/\gamma_k^{mol}$ multiplies a potential energy consisting of two distinct contributions, namely, a nuclear–electronic and a nuclear–nuclear part. Now, one can use two independent multipliers, $1/\gamma_k^{el}$ and $1/\gamma_k^{nucl}$, and write

$$E_k = \frac{1}{\gamma_k^{el}} V_{ne,k} + \frac{1}{\gamma_k^{nucl}} Z_k \sum_{l \neq k} \frac{Z_l}{R_{kl}} \tag{6.2}$$

by letting

$$\frac{1}{\gamma_k^{mol}} = \left(\frac{1}{\gamma_k^{el}} V_{ne,k} + \frac{1}{\gamma_k^{nucl}} Z_k \sum_{l \neq k} \frac{Z_l}{R_{kl}} \right) \frac{1}{V_k} \tag{6.3}$$

i.e., by taking $1/\gamma_k^{mol}$ as the weighted average of $1/\gamma_k^{el}$ (for the nuclear–electronic part) and $1/\gamma_k^{nucl}$ (for the nuclear–nuclear part). Detailed SCF computations indicate that $\gamma_k^{el} = \gamma_k^{nucl}$, at least within the precision permitted by this type of analysis. This result settles an important question regarding the constancy of the γ_k^{mol} parameter introduced in the defining equation (2.15). Indeed, if it were $\gamma_k^{el} \neq \gamma_k^{nucl}$, then γ_k^{mol} would change to some extent from molecule to molecule, depending on the relative weights of $V_{ne,k}$ and $Z_k \sum_{l \neq k} Z_l/R_{kl}$ in eq. (6.3). The results are included in Table 6.1.

In closing, let us examine the γ values for molecules using (2.16). It is clear that the average $1/\gamma$ depends on the γ_k^{mol} values of Table 6.1 and on the weights of the potential energies V_k of the individual atoms. Now, the weight of the heavy atoms, of the order of -88 au for carbon and -175 au for oxygen, for example, is considerably larger than that of hydrogen, ~ -1 au. As a consequence, the final result for γ is in most cases close to $\frac{7}{3}$. This is a well-known fact [3–8]: SCF calculations made for a great variety of organic molecules indicate that $\gamma \simeq 7/3$ within approximately $\pm 1\%$. We use this approximation for γ^v, keeping in mind that it affects only a small portion of the total atomization energy that shall be calculated.

The calculation of $(\partial E_k^{vs}/\partial N_k)^\circ$ and $(\partial^2 E_k^{vs}/\partial N_k^2)^\circ$

We now direct our attention to the calculation of the a_{kl} parameters. The appropriate first and second derivatives, $(\partial E_k^{vs}/\partial N_k)^\circ$ and $(\partial^2 E_k^{vs}/\partial N_k^2)^\circ$, are conveniently obtained from SCF–Xα theory [9], which offers the definite advantage of permitting calculations for any desired integer and fractional electron population. It is, indeed, important to account for the fact that these derivatives depend on N_k. The difficulty is that this type of calculations cannot be performed directly on atoms which are actually part of a molecule. One resorts to model free atom calculations to mimic the behavior of atoms which are in a molecule but do not experience interactions with the other atoms in the molecule.

For hydrogen $\alpha = 0.686$ is appropriate for a partially negative atom (like that found in ethane) and reproduces correctly its electron affinity. For $N_H = 1.0117$ e, it is $(\partial E_H/\partial N_H)^\circ = -0.195$ au. For the carbon atoms of saturated hydrocarbons it must be considered, first, that optimized SCF computations indicate that any gain in electronic charge, with respect to the ethane carbon, occurs at the $2s$ level [3]. Second, $(9s5p|6s) \rightarrow [5s3p|3s]$ calculations of methane and ethane, using Dunning's exponents [10] and optimum contraction vectors, indicate $2s$ populations of 1.42–1.46 e. Finally, SCF–Xα computations give $(\partial E_k/\partial N_k)$ values of -20.29, -19.87 and -19.26 eV for $2s$ populations of 1.40, 1.45 and 1.50 e, respectively, by using the $\alpha = 0.75928$ value recommended by Schwarz [11]. These results suggest that the appropriate $(\partial E_k/\partial N_k)$ derivative can be reasonably estimated at -0.735 au $(-20$ eV$)$ for sp^3 carbon atoms. The results for sp^2 carbon atoms [12] as well as those for oxygen [3] follow from similar SCF–Xα calculations (Table 6.2).

Turning now to the other terms occurring in (3.35), we approximate

TABLE 6.2. Selected first and second energy derivatives, au

Atom	Orbital	Popul.	$(\partial E_k/\partial N_k)^\circ$	$(\partial^2 E_k/\partial N_k^2)^\circ$
H	$1s$	1.0117	-0.195	
C	$2s$	1.44	-0.735	0.45
C	$2p$	1.97	-0.200	0.37
C	σ	3.00	-0.375	
C (C_2H_4)	π	1.00	-0.246	
C (C_6H_6)	π	1.00	-0.262	
O	$2p$	4.023	-0.316	0.50
O	$2p$	3.995	-0.331	

γ^v by $\frac{7}{3}$ and use the appropriate γ_k^{mol}'s of Table 6.1. The $\langle r_{kl}^{-1}\rangle$ terms are approximated by the inverse internuclear distances, R_{kl}^{-1}. For the $C(sp^3)$–H bonds, for example, assuming $R_{CH} = 1.08$ Å, eq. (3.35) gives $a_{HC} = -1.007$ and $a_{CH} = -0.394$ atomic units. For the $C(sp^3)$–$C(sp^3)$ bonds, assuming $R_{CC} = 1.53$ Å, it is $a_{CC} = -0.777$ au. We prefer kcal mol^{-1}me^{-1} units for these parameters. Using the conversion factor 1 hartree = 627.51 kcal mol^{-1}, we get $a_{HC} = -0.632$, $a_{CH} = -0.247$ and $a_{CC} = -0.488$ kcal mol^{-1}me^{-1}. These are the standard a_{kl} values [13], and the only ones that are required, for the saturated hydrocarbons[2]. Their bond energies are thus given by

$$\varepsilon_{CH} = \varepsilon_{CH}^\circ - 0.247\Delta q_C - 0.632\Delta q_H \quad \text{kcal mol}^{-1} \quad (6.4)$$

$$\varepsilon_{CC} = \varepsilon_{CC}^\circ - 0.488\Delta q_{C_k} - 0.488\Delta q_{C_l} \quad \text{kcal mol}^{-1} \quad (6.5)$$

where the charges are expressed in me (10^{-3} e) units. These expressions certainly show that small charge variations significantly affect bond energies.

A multitude of a_{CH} and a_{CC} parameters occur in the study of olefins and benzenoid hydrocarbons. It is unpractical to list them here. It is quicker to calculate them when required. For the bonds formed by carbon atoms with $l = C$ or H, it suffices to use the following obvious approximation of (3.35)

$$a_{Cl} = \frac{1}{v}\left(\frac{\partial E}{\partial N}\right)^\circ - \frac{3}{7}\cdot\frac{Z_l^{eff}}{R_{Cl}} \quad (6.6)$$

and to select the appropriate derivatives from Table 6.2.

---

[2]There are, of course, minor and unfortunately unavoidable imprecisions associated with all the parameters entering a_{kl}. Evidently, they are difficult to assess but, in this selection of a_{kl} results, they seem to compensate one for another (Section 5).

6.3   Basic functions

This Section describes applications of the formulas developed in Chapter 3.6
for the centroids of valence atomic orbitals, namely the use of the function
F, eq. (3.55). Then we apply these results and construct reference bond
energies of particular interest for the study of unsaturated hydrocarbons.

Geometry changes and displacements of charge centroids

The problems to be solved are best illustrated by a typical example. Take ε_1°
as the carbon–carbon bond energy in ethane, with $R_{CC}^\circ = 1.531$ Å; calculate
the energy of a $C(sp^3)$–$C(sp^2)$ bond like that found in olefins. The latter is
for reference charges which are: 35.1 me for the sp^3 carbon (like in ethane)
and 7.7 me for the sp^2 carbon (like in ethylene). This transformation is
schematically represented in Figure 6.1.

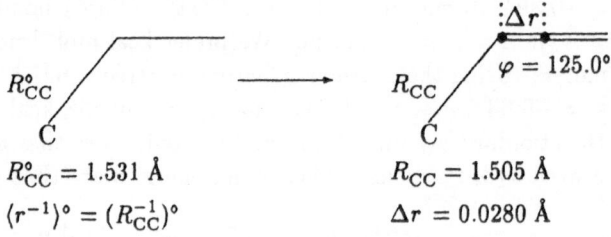

FIGURE 6.1. The bond transformation $C(sp^3)$–$C(sp^3) \longrightarrow C(sp^3)$–$C(sp^2)$

This transformation involves i) a change of charge, ii) a change of the bond
length and iii) the displacement, Δr, of the π-orbital centroid, on the C=C
axis, with respect to the nuclear position of the sp^2 atom (identified here as
atom l). The change of charge is dealt with in the next Subsection. Here we
concentrate on topics ii) and iii) with the help of eqs. (3.53)–(3.55).

Let us first rewrite (3.55) as follows

$$F_{kl} = -\frac{3}{7} Z_k^{\text{eff}} Z_l^{\text{eff}} \left[R_{kl}^{-1} - \left(R_{kl}^{-1} \right)^\circ - \left(\langle r_{kl}^{-1} \rangle - \langle r_{kl}^{-1} \rangle^\circ \right) \right]$$
$$-\frac{3}{7} Z_k^{\text{eff}} q_l \left(\langle r_{kl}^{-1} \rangle - \langle r_{kl}^{-1} \rangle^\circ \right). \tag{6.7}$$

Attention must be given to the indices. F_{kl} mirrors, so to speak, how nucleus
k 'sees' the electrons of atom l, but F_{lk} if for nucleus l in the field of the
electrons of atom k. The sum $F_{kl} + F_{lk}$ gives the total F for that k–l bond.

The relevant bond distance, R_{kl}°, of the initial molecule and the geometry of the final product, i.e., R and φ (see Figure 6.1), are known from the problem. The actual calculations involve four steps: i) calculate Δr as indicated in Chapter 3.6 and ii), applying the theorem (3.54), calculate the distance between the centroid of the π orbital on atom l and nucleus k; iii) then use the monopole approximation (3.53) and obtain $\langle r_{kl}^{-1} \rangle$, the average distance between the valence electrons of l and nucleus k; finally iv) use this $\langle r_{kl}^{-1} \rangle$ result in eq. (6.7) and find F_{kl}. $\langle r_{kl}^{-1} \rangle^\circ$ is usually known from the context of the problem [as in the example given in Figure 6.1, where $\langle r_{kl}^{-1} \rangle^\circ = (R_{kl}^{-1})^\circ$] or must be calculated following steps i)–iii). Note that step iii) involves the total valence electron population of atom l, namely 3.9923 e for the carbon atom of ethylene, whose net charge is 7.7 me. For sp^3 carbon atoms and those of aromatic rings, q_l is 35.1 or 13.2 me, respectively. Steps ii)–iv) are easy to carry out by hand. As regards step i) we approach the problem by observing the following guidelines suggested by direct computations of Δr.

For σ systems (sp^3 carbons and hydrogen), the charges are taken at their respective nuclear positions. This is part of the monopole approximation (3.53). For the aryl carbon atoms we have $\Delta r = 0$ for obvious symmetry reasons. This leaves us with the olefinic double bonds. They are dealt with in the following manner.

The basic idea is to find representative $C(sp^2)$–$C(sp^3)$ and $C(sp^2)$–H bonds for general use in applications to olefins, i.e., to find an acceptable approximation that bypasses lengthy computations of Δr for each molecule of interest. Let us first evaluate the risks. Consider the $C(sp^2)$–$C(sp^3)$ bond of tetramethylethylene, with $\Delta r = 0.0280$ Å, formed from that of ethane (Figure 6.1) with $F = -4.30$ kcal mol^{-1} (details are given in example 1). Consider also a standard $C(sp^2)$–H bond, calculated for $\Delta r = 0.0290$ Å and $\varphi = 121.7°$, formed from that of ethane with $F = -1.87$ kcal mol^{-1} (example 2). Now consider propene: on the side of the sp^2 carbon carrying the methyl substituent, direct calculations indicate that $\Delta r = 0.0299$ Å. This $C(sp^2)$–$C(sp^3)$ bond is formed from that of ethane with $F = -4.53$ kcal mol^{-1}, which is 0.23 kcal mol^{-1} *more* negative than that of tetramethylethylene. At the other end of the double bond of propene, in turn, the displacement is reduced to 0.0281Å and the total F of the two hydrogens at that end is -3.53 kcal mol^{-1}. For the third CH bond, i.e., that formed by the carbon bearing the methyl group, F is -1.76 kcal mol^{-1}. The total F for the three CH bonds is thus -5.29 kcal mol^{-1}, i.e., 0.32 kcal mol^{-1} *less* negative than three times the assumed reference, $3(-1.87) = -5.61$ kcal mol^{-1}. Briefly, the price payed for this approximation is a discrepancy of 0.09 kcal mol^{-1}

between the approximated and the direct results obtained for propene. The selection of a standard value of 0.029 Å for the calculation of $C(sp^2)$–H bonds seems reasonable—that of ethylene itself is 0.0292 Å—as well as the selection of the symmetrical tetramethylethylene molecule, with $\Delta r = 0.0280$ Å, for the evaluation of a model $C(sp^2)$–$C(sp^3)$ bond. These simplifying rules are applied in all the forthcoming calculations of olefins.

Let us now work out some examples. The energies are given in kcal mol^{-1} using the conversion factor 1 hartree = 627.51 kcal mol^{-1}. Distances and inverse distances are indicated in Å and Å^{-1}, respectively, with 1 bohr = 0.52917 Å. The angles φ are reported in degrees. The calculations are made for reference charges, i.e., 35.1, 7.7 and 13.2 me for sp^3, ethylenic and aryl carbons, respectively.

Example 1. We begin with the CC bond of ethane, $R^\circ_{kl} = 1.531$, assuming $\langle r^{-1}_{kl} \rangle^\circ = (R^{-1}_{kl})^\circ$ for this σ system. This bond is tranformed into the $C(sp^3)$–$C(sp^2)$ bond of tetramethylethylene, with $R = 1.505$, $\varphi = 125.0$ (experimental geometry) and $\Delta r = 0.0280$ (from a direct computation of Δr). The cosine theorem (3.54) tells us that the centroid of the π orbital is at a distance of 1.521233 from the nucleus of the sp^3 carbon. Using this result in (3.53) we get $\langle r^{-1}_{kl} \rangle = 0.66268$. Finally, eq. (6.7) gives $F_{kl} = -4.076$. For the calculation of F_{lk} (i.e., for the electrons of the methyl carbon in the field of the nuclear charge of the sp^2 carbon), we have $\langle r^{-1}_{lk} \rangle^\circ = (R^{-1}_{lk})^\circ$, $\langle r^{-1}_{lk} \rangle = R^{-1}_{lk}$ and $F_{lk} = -0.225$. The total F for this transformation is thus $F = -4.30$.

Example 2. We transform a H–$C(sp^3)$ bond, $R^\circ_{kl} = 1.08$, into a H–$C(sp^2)$ bond, $R_{kl} = 1.08$, with $\Delta r = 0.029$. (The sp^2 carbon is atom l.) For ethylene and, more generally, for CH bonds in *trans*-olefins, $\varphi \simeq 121.7$. Using this angle we get $\langle r^{-1}_{kl} \rangle = 0.92264$. For the original CH bond it is $\langle r^{-1}_{kl} \rangle^\circ = (R^{-1}_{kl})^\circ$. Thus we obtain $F_{kl} = F = -1.87$ because $F_{lk} = 0$. For *cis*-olefins, we use $\varphi = 117.4$, the angle calculated for *cis*-butene, and get $F = -1.67$.

Example 3. To make things a little more interesting, we transform the $C(sp^3)$–$C(sp^2)$ bond of tetramethylethylene (example 1) into a $C(sp^2)$–Csp^2 single bond like that found in 1,3-butadiene, assuming $R = 1.49$, $\varphi = 123$ and $\Delta r = 0.028$. The latter three entries give $\langle r^{-1}_{kl} \rangle = 0.66942$. Now we calculate F_{kl}, where l is the sp^2 carbon of example 1; thus we use here $R^\circ = 1.505$ and $\langle r^{-1}_{kl} \rangle^\circ = 0.66268$. Eq. (6.7) gives $F_{kl} = 0.086$. F_{lk} is for the sp^3 carbon of the original bond that is transformed into an sp^2 carbon, again with $\langle r^{-1}_{lk} \rangle = 0.66942$, but with $\langle r^{-1}_{lk} \rangle^\circ = (R^{-1}_{lk})^\circ$. So we obtain $F_{lk} = -3.940$ and $F = F_{kl} + F_{lk} = -3.85$.

Example 4. Here we use the $C(sp^3)$–$C(sp^2)$ bond of tetramethylethylene and transform it into a $C(Ar)$–$C(sp^2)$ bond like that found in styrene, $R = 1.445$, $\varphi = 126$ (calculated geometry) with $\Delta r = 0.028$. Let l be the $C(sp^2)$ atom of the original bond, with $R^\circ_{kl} = 1.505$ and $\langle r^{-1}_{kl} \rangle^\circ = 0.66268$ (see example 1). For styrene we get $\langle r^{-1}_{kl} \rangle = 0.69007$ and $F_{kl} = -0.575$. For the aryl carbon atom and its electrons it is $\langle r^{-1}_{lk} \rangle^\circ = (R^{-1}_{lk})^\circ$ and $\langle r^{-1}_{lk} \rangle = R^{-1}_{lk}$. Remenbering that $q_k = 13.2 \times 10^{-3}$ e, we get $F_{lk} = -0.207$ and, finally, $F = -0.78$ for this transformation.

It is clear that some information is lost as a result of our approximation, namely as regards the individual bonds formed by sp^2 carbon atoms, but the total F values are expected to be generally reasonably accurate. The salient feature of ethylenic double bonds, i.e., the inward displacement of π-orbital centroids on the C=C axis revealed by direct calculations, and its important role in energy calculations, can now be put in a clear perspective and efficiently tested for large collections of molecules.

Reference bond energies

The description of saturated, olefinic and benzenoid hydrocarbons involves four basic reference bond energies. They are: ε_1^o for the CC bond of ethane, ε_2^o for the double bond of ethylene, ε_3^o for the CC bonds of benzene and ε_{10}^o for the CH bond of ethane. The description of ethers also involves a C–O reference energy and a C=O bond energy is required for the carbonyl compounds. The energies of carbon–carbon single bonds involving all possible combinations of sp^3, olefinic sp^2 and aryl sp^2 carbons are deduced from ε_1^o. The CH bond energies involving sp^2 carbons are generated from ε_{10}^o. The basic reference bond energies are, in principle, given by eq. (3.18). Accurate applications of this equation are difficult, however.

Approximate solutions can be found in some favorable cases with the help of (3.15), as indicated in the following example worked out for ethane.

Example 5. First we apply (2.5). SCF results for methane and ethane are[3]: $V_C^{CH_4} = -88.52015$, $V_H^{CH_4} = -1.12634$, $V_C^{C_2H_6} = -88.45999$ and $V_H^{C_2H_6} = -1.13355$ au, with total energies $E^{mol} = -40.2090$ and -79.2513 au, respectively. Next we apply $E^{mol} = \sum_k (1/\gamma_k) V_k$ [from (2.15)–(2.17)] assuming that the SCF γ_k^{mol}'s are valid for real molecules and rescale the SCF potential energies to reproduce the corresponding experimental energies of atomization; so we get the *rescaled values* $V_C^{CH_4} = -89.2305$, $V_H^{CH_4} = -1.1354$, $V_C^{C_2H_6} = -89.1270$ and $V_H^{C_2H_6} = -1.1421$ au. At last we can estimate the CH and CC bond energies of ethane. This can be done in two ways. *Method 1.* Remembering that $V_{ne} = -1$ au for the H atom, eq. (3.15) tells us that, to a good approximation, $(\partial \varepsilon_{CH}/\partial Z_H)$ is 0.1354 au for methane and 0.1421 au for ethane. Consider now (3.18) and suppose that $(\partial \varepsilon_{CH}/\partial Z_C)$ is at least approximately the same in both molecules[4]. In this approximation, eq. (3.18) tells us that the ε_{CH} energy of ethane is larger than that of methane by $\sim \frac{1}{2}(0.1421 - 0.1354)$ au $= 2.10$ kcal mol^{-1}. Since the latter is ~ 104.81 kcal mol^{-1}, we find for ethane that $\varepsilon_{CH}^o \simeq 106.91$ and thus $\varepsilon_{CC}^o \simeq 69.1$

---

[3] Pople's 6-311G** basis [14] and experimental geometries [15] of methane ($R_{CH} = 1.085$ Å) and ethane ($R_{CC} = 1.531$ Å, $R_{CH} = 1.096$ Å and $\angle HCH = 107.8°$) were used [16].

[4] This unavoidable hypothesis, under the present circumstances, seems justifiable on counts that the relative charge difference between these carbon atoms is very small.

kcal mol^{-1}, by reference to the experimental energies of methane and ethane[5]
Method 2. Here we use $V_C^{at} = -88.5307$ au for the carbon atom, deduced from
its experimental energy, -37.8558 au, and the γ_C^{at} of Table 6.1. Equation (3.15)
gives, in atomic units, $-89.2305 = -88.5307 - Z_C[4(\partial\varepsilon_{CH}/\partial Z_C)]$ for methane
and $-89.1270 = -88.5307 - Z_C[3(\partial\varepsilon_{CH}/\partial Z_C) + (\partial\varepsilon_{CC}/\partial Z_C)]$ for ethane. Now,
assuming as before the same value of $(\partial\varepsilon_{CH}/\partial Z_C)$ for the two molecules, we get
$(\partial\varepsilon_{CH}/\partial Z_C) = 0.02916$ and $(\partial\varepsilon_{CC}/\partial Z_C) = 0.0119$ au. So we obtain from (3.18)
that $\varepsilon_{CC} = 69.0$ kcal mol^{-1}. Note that the first calculation uses only the differ-
ence between the potential energies $V_H^{C_2H_6}$ and $V_H^{CH_4}$ to get the CH bond energy
of ethane from that of methane, which is known. The second calculation involves
$V_C^{C_2H_6}$, $V_C^{CH_4}$ and V_C^{atom} *and*, most importantly, it depends heavily on the γ parameters required for solving (3.18). Under these difficult circumstances, it is gratifying
that both methods give approximately the same result for ε_{CC}°. Similar calculations,
using exclusively SCF input data, indicate that the CC bond energies of ethylene
and benzene are, respectively, ~ 2.0 and ~ 1.65 times that of ethane [16], nicely
supporting the empirically adjusted ε_2° and ε_3° results propounded in Table 6.3.

The calculation of ε_{CH}° and ε_{CC}° using (3.18) suffers from imprecisions and
can do no better than suggesting an approximate value, ~ 69 kcal mol^{-1}, for
the CC bond energy of ethane. The final selection, described in Section 5,
is $\varepsilon_1^\circ = 69.633$ kcal mol^{-1} for the CC bond of ethane and $\varepsilon_{10}^\circ = 106.806$
kcal mol^{-1} for its CH bonds. These values shall be used for the saturated
hydrocarbons. Here we use them to get the CC and CH bonds occurring
in unsaturated hydrocarbons by considering both the contribution of F, eq.
(6.7), and the change of charge given by $a_{lk}\Delta q_l$, eq. (3.39). The new reference
bonds thus obtained are indicated in Table 6.3.

The appropriate transformations are described in examples 6–12. The units are
those used for examples 1–4. The a_{lk} parameters, in kcal mol^{-1}me^{-1} units, stand
for the interatomic distance indicated in parentheses, in Å. This distance is that
of the *initial* bond because the change in geometry is part of F. The derivative
$(\partial E_l/\partial N_l)^\circ = -0.375$ au, on the other hand, is that of the newly formed sp^2
carbon ($=$ atom l). This approach implicitly maximizes the association of all energy
changes resulting from the rehybridization of a carbon atom with the multiple bond
created during this rehybridization. In this energy partitioning scheme, the gain in
electronic charge, e.g., -27.4 me in going from the ethane to the ethylene carbon
atom, is considered to take place at the σ level: the fact that a π orbital 'separates
out' with a concurrent change of the atom's own energy is regarded as part of the
newly formed multiple bond and its energy. The a_{lk} parameters are easily obtained
from eq. (6.6) remembering that $v = 3$ for the sp^2 carbon atoms.

Example 6. ε_4° is obtained from ε_1° as: $\varepsilon_4^\circ = \varepsilon_1^\circ + a_{CC}(7.7 - 35.1) + F$, with
$a_{CC}(1.531) = -0.450_3$ and $F = -4.30$, calculated in example 1.

---

[5]The experimental atomization energies are $\Delta E_a^* = 419.24$ and 710.54 kcal mol^{-1} for
methane and ethane, respectively.

TABLE 6.3. Selected reference bonds and their energies, kcal mol^{-1}

Bond	Type	Occurrence	R_{kl}, Å	q_k^o, me	q_l^o, me	Energy
ε_1^o	$C(sp^3)$-$C(sp^3)$	Ethane	1.531	35.1	35.1	69.63
ε_2^o	$C(sp^2)$=$C(sp^2)$	Ethene	1.34	7.7	7.7	139.37
ε_3^o	$C(Ar)$:::$C(Ar)$	Benzene	1.397	13.2	13.2	115.39
ε_4^o	$C(sp^3)$-$C(sp^2)$	Olefins	1.505	35.1	7.7	77.67
ε_5^o	$C(sp^2)$-$C(sp^2)$	Butadiene	1.49	7.7	7.7	89.14
ε_6^o	$C(Ar)$-$C(sp^3)$	Toluene	1.48	13.2	35.1	79.33
ε_7^o	$C(Ar)$-$C(sp^2)$	Styrene	1.445	13.2	7.7	89.69
ε_8^o	$C(Ar)$-$C(Ar)$	Aromatics, conj.	1.397	13.2	13.2	91.21
ε_9^o	$C(Ar)$-$C(Ar)$	Biphenyl	1.49	13.2	13.2	89.03
ε_{10}^o	$C(sp^3)$-H	Ethane	1.08	35.1	-11.7	106.81
ε_{11}^o	$C(sp^2)$-H	Ethene	1.08	7.7	-11.7	110.69
ε_{12}^o	$C(Ar)$-H	Benzene	1.08	13.2	-11.7	111.41

Example 7. ε_5^o follows from ε_4^o, i.e., $\varepsilon_5^o = \varepsilon_4^o + a_{CC}(7.7 - 35.1) + F +$ conjug., with $a_{CC}(1.505) = -0.456_7$ and $F = -3.85$ (see example 3). Conjugation is accounted for as indicated in Chapter 3.8 and tentatively estimated at 2.8 kcal mol^{-1}.

Example 8. The $C(Ar)$-$C(sp^3)$ bond like that found in toluene is given by $\varepsilon_6^o = \varepsilon_1^o + a_{CC}(13.2 - 35.1) + F$, with $a_{CC}(1.531) = -0.450_3$. F is calculated by observing that $\langle r_{kl}^{-1} \rangle^o = (R_{kl}^{-1})^o$ for ethane and $\langle r_{kl}^{-1} \rangle = R_{kl}^{-1}$ for toluene, with $R_{CC} = 1.48$ (calculated geometry), which gives $F = -0.169$ and $\varepsilon_6^o = 79.33$.

Example 9. The conjugated carbon–carbon single bond like that of styrene is deduced from ε_4^o as $\varepsilon_7^o = \varepsilon_4^o + a_{CC}(13.2 - 35.1) + F +$ conjug., where $a_{CC}(1.505) = -0.456_7$. $F = -0.78$, as shown in example 4, and conjugation adds 2.8 kcal mol^{-1}.

Example 10. ε_8^o is for a conjugated carbon–carbon single bond formed by two aryl sp^2 carbon atoms, at a distance 1.397 (which is the experimental distance in benzene [17]). This bond is calculated from ε_1^o, that is $\varepsilon_8^o = \varepsilon_1^o + 2a_{CC}(13.2 - 35.1) + F +$ conjug., with $a_{CC}(1.531) = -0.450_3$. $F = -0.94_2$ is obtained by observing that $\langle r_{kl}^{-1} \rangle^o = (R_{kl}^{-1})^o$ and $\langle r_{kl}^{-1} \rangle = R_{kl}^{-1}$, with $R_{kl}^o = 1.531$ and $R_{kl} = 1.397$.

Example 11. ε_9^o stands for a nonconjugated carbon–carbon bond between aryl carbon atoms, like that occurring in biphenyl, with $R = 1.49$ [18]. Remembering that ε_6^o was obtained from ε_1^o by replacing one CH$_3$ of ethane by a phenyl group, the central CC bond of biphenyl is approximated as $2(\varepsilon_6^o - \varepsilon_1^o) + \varepsilon_1 = 89.03$.

Example 12. The reference energies of $C(sp^2)$-H bonds are readily obtained from ε_{10}^o, namely, $\varepsilon_{11}^o = \varepsilon_{10}^o + a_{CH}(7.7 - 35.1) + F$, with $a_{CH}(1.08) = -0.210_2$ and $F = -1.87$ (example 2), whereas $F = 0$ for $\varepsilon_{12}^o = \varepsilon_{10}^o + a_{CH}(13.2 - 35.1) = 111.41$.

The bond energies reported in Table 6.3 correspond to the conditions specified in examples 1–12. They could possibly be bettered for general use, but we take them as indicated.

6.4   Nonbonded interactions

The idea behind this brief survey of nonbonded interactions is to find an acceptable approximation permitting to get rid of them as explicit terms requiring separate calculations, e.g., using eq. (3.11). We shall examine to what extent nonbonded Coulomb-type interactions are at least approximately additive. The formulation of additivity is presented here for $C_nH_{2n+2-2m}$ hydrocarbons [19].

Let X be a molecular property (e.g., E_{nb}) and $X^{C_2H_6}$, X^{CH_4} the corresponding values for ethane and methane, respectively. If X is exactly addidive, then

$$X = (1 - m)X^{C_2H_6} + (n - 2 + 2m)\left(X^{C_2H_6} - X^{CH_4}\right) \qquad (6.8)$$

where $X^{C_2H_6} - X^{CH_4}$ is the change in X on going from methane to ethane, i.e., the contribution of one CH_2 group. The meaning of (6.8) is obvious for noncyclic molecules ($m = 0$). For example, the X value for propane is that of ethane plus the increment corresponding to one added CH_2 group. For cyclohexane ($m = 1$) which consists of $n - 2 + 2m = 6$ CH_2 groups, the $(1 - m)X^{C_2H_6}$ term of (6.8) cancels. Decalin is constructed from two cyclohexane units. In this case $(n - 2 + 2m)(X^{C_2H_6} - X^{CH_4})$ accounts for 12 CH_2 groups, but one additional $X^{C_2H_6}$ contribution (i.e., that of two CH_2 and two H atoms) is subtracted with respect to cyclohexane, i.e., a total of two $X^{C_2H_6}$ contributions with respect to noncyclic alkanes. Similar arguments applied to other polycyclic saturated hydrocarbons verify the validity of (6.8) as a formulation for exact additivity.

It turns out [19] that the nonbonded terms of saturated hydrocarbons are approximately additive, i.e.,

$$E_{nb} \simeq (1 - m)E_{nb}^{C_2H_6} + (n - 2 + 2m)\left(E_{nb}^{C_2H_6} - E_{nb}^{CH_4}\right) \qquad (6.9)$$

if we agree upon accepting errors of $\sim$0.1 kcal mol^{-1} due to the neglect of differences between isomers. Branching causes a systematic trend towards larger E_{nb}'s but situations of extreme steric crowding, such as those encountered in 2,2,3,3-tetramethylbutane and, to a lesser extent, in 2,2,3-trimethylbutane, are required in order to produce sizeable departures from quasi-additivity, eq. (6.9). With these reservations in mind, a simple way of taking advantage of the quasi-additivity of nonbonded interactions in saturated hydrocarbons shall be implemented in Section 6.5. The unimportant loss in precision is largely justified by the considerable simplification thus achieved in calculations of atomization energies.

6.5 Saturated hydrocarbons

Here we examine saturated hydrocarbons, $C_nH_{2n+2-2m}$, containing n carbon atoms and m chair or boat six-membered cycles. Their general energy formula was deduced in Chapter 3.5. Using these results, namely eqs. (3.40) and (3.42), as well as (3.4), equation (3.3) becomes

$$\Delta E_a^* = (n - 1 + m)\varepsilon_{CC}^o + (2n + 2 - 2m)\varepsilon_{CH}^o + A_1 \sum N_{CC}\Delta q_C$$
$$+ A_2 \sum \Delta q_C + (n - 2 + 2m)a_{HC}\, q_H^o - E_{nb} \qquad (6.10)$$

where $A_1 = a_{CC} - a_{CH}$ and $A_2 = 4a_{CH} - a_{HC}$. N_{CC} is the number of CC bonds formed by the carbon whose net charge differs by Δq_C from that of the ethane carbon, q_C^o. E_{nb} is calculated with the help of (3.11). In principle we know all the quantities required for solving (6.10) but shall momentarily suppose that this is not the case and submit eq. (6.10) to three instructive tests using experimental ΔE_a^* results.

The first test concerns the relative ordering of the carbon net charges. We write eq. (6.10) for a representative selection of molecules using explicitly the net charges given by eqs. (5.7) and (5.8), i.e.,

$$q_C = q_C^{Mulliken} + N_{CH}\, p$$
$$q_H = q_H^{Mulliken} - p$$

and determine the unknown p by a least-square analysis. For fully optimized STO-3G charges (Chapter 5.5) one obtains [13] $p = (30.3 \pm 0.3) \times 10^{-3}$ e. Remember that the same set of Mulliken charges, when compared to the corresponding ^{13}C NMR shifts, yields $p = 30.12 \times 10^{-3}$ e and gives $\delta_C = -237.1\,\Delta q/q_C^o$ ppm relative to ethane. This test shows[6] that the same definition of charges satisfies both eq. (6.10) and the relationship between δ_C and Δq_C. So we can safely use the latter for deriving the required Δq_C's—a circumstance greatly facilitating applications of eq. (6.10)—and write (with the appropriate unit transformations) $A_1 \sum N_{CC}\, \delta_C$ instead of $A_1 \sum N_{CC}\Delta q_C$ and $A_2 \sum \delta_C$ instead of $A_2 \sum \Delta q_C$.

The second test concerns the parameters A_1 and A_2. They are readily obtained from least-square fittings of experimental ΔE_a^*'s with eq. (6.10). The empirical ratio $A_2/A_1 = 1.486$ thus determined equals that anticipated from the theoretical values, $a_{CC} = -0.777$, $a_{CH} = -0.394$ and $a_{HC} = -1.007$

---

[6]This sort of analysis holds independently of the LCAO-MO method selected for calculating Mulliken charges (Chapter 5.5).

au, described in Section 2. So we have $A_1 = -0.383$ and $A_2 = -0.569$ au which are conveniently rewritten in kcal mol^{-1} ppm^{-1} units as $A_1 = 0.383(627.51/237.1)q_C^o$ and $A_2 = 0.569(627.51/237.1)q_C^o$.

The final test uses these A_1 and A_2 parameters in applications of eq. (6.10). Solving (6.10) for ethane, we find for its bonded part

$$\Delta E_{a,bd}^{*C_2H_6} = \Delta E_a^{*C_2H_6} + E_{nb}^{C_2H_6}$$

that $\Delta E_{a,bd}^{*C_2H_6} = \varepsilon_{CC}^o + 6\varepsilon_{CH}^o$. Similarly, we get for methane that $\Delta E_{a,bd}^{*CH_4} = 4\varepsilon_{CH}^o + A_2\delta_C^{CH_4} - a_{HC}\, q_H^o$, where $\delta_C^{CH_4} (= -8$ ppm) is the NMR shift of the methane carbon relative to ethane. Now we rearrange eq. (6.10) as follows

$$\begin{aligned}
2n\varepsilon_{CC}^o = {} & \Delta E_a^* + E_{nb} + n\left(\Delta E_{a,bd}^{*C_2H_6} - 2\Delta E_{a,bd}^{*CH_4} + 2A_2\delta_C^{CH_4} + a_{HC}\, q_C^o\right) \\
& + (1-m)\left(\Delta E_{a,bd}^{*C_2H_6} - 2\Delta E_{a,bd}^{*CH_4} + 2A_2\delta_C^{CH_4}\right) \\
& - A_1 \sum N_{CC}\delta_C - A_2 \sum \delta_C
\end{aligned} \qquad (6.11)$$

and take advantage of the fact that ε_{CC}^o is constant by definition. Note the presence of q_C^o in the right-hand side of (6.11), namely in A_1 and A_2. Application of (6.11) to a series of molecules indicates that the constraint of a constant ε_{CC}^o is satisfied for $q_C^o \simeq 0.035$ e, with $\varepsilon_{CC}^o \simeq 69.7$ kcal mol^{-1} [13]. Our optimum choice, $q_C^o = 35.1$ me, yields $A_1 = 0.0356$ and $A_2 = 0.0529$ kcal mol^{-1} ppm^{-1} and $\varepsilon_{CC}^o = 69.633$ kcal mol^{-1}. This choice remains open to discussion as regards its precision but there is presently no way we could fairly discriminate between these results and those predicted by theory (Table 6.4). At this level, arguments are limited by both experimental and theoretical uncertainties and revolve only about minor (≤ 0.15 kcal mol^{-1}) statistical improvements of the agreement between calculated and experimental atomization energies.

TABLE 6.4. Verification of eq. (6.10)

Parameter	Empirical	Theoretical
A_2/A_1	1.486	1.486
q_C^o(ethane), me	35.1	37.8
ε_{CC}^o, kcal mol^{-1}	69.633	~ 69

It is fair to conclude that these tests demonstrate the general validity of the theory embodied in (6.10). The theories underlying energy calculations

TABLE 6.5. Comparison between calculated and experimental atomization energies of saturated hydrocarbons, kcal mol^{-1}

Molecule	$\sum_{k<l} \varepsilon_{kl}^o$	$\sum_k \sum_l a_{kl} \Delta q_k$	E_{nb}	ΔE_a^* Calcd.	ΔE_a^* Exptl.
Methane	427.22	−7.82	0.09	419.31	419.24
Ethane	710.47	0.00	−0.07	710.54	710.54
Propane	993.71	10.38	−0.20	1004.29	1004.07
Butane	1276.96	20.82	−0.32	1298.10	1298.15
Isobutane	1276.96	22.78	−0.28	1300.02	1299.70
Pentane	1560.20	31.25	−0.44	1591.90	1592.18
Isopentane	1560.20	32.58	−0.39	1593.17	1593.43
Neopentane	1560.20	35.54	−0.32	1596.06	1595.94
Hexane	1843.45	41.69	−0.55	1885.69	1885.95
2-Methylpentane	1843.45	43.05	−0.50	1887.00	1886.86
3-Methylpentane	1843.45	42.38	−0.50	1886.33	1886.27
2,2-Dimethylbutane	1843.45	44.53	−0.43	1888.41	1888.86
2,3-Dimethylbutane	1843.45	43.32	−0.39	1887.16	1887.14
2,2,3-Trimethylbutane	2126.69	55.00	−0.46	2182.15	2181.90
Cyclohexane	1699.47	60.60	−0.73	1760.80	1760.82
Methylcyclohexane	1982.72	79.19	−0.83	2056.74	2057.13
trans-Decalin	2688.47	125.83	−1.26	2815.56	2815.50
Adamantane	2544.49	142.18	−1.31	2687.98	2688.05
Bicyclo[2.2.2]octane	2121.98	95.50	−1.03	2218.51	2218.40

and the characterization of physically meaningful charges are entirely independent of one another. Here we have learned that the charges satisfying the constraints described in Chapter 5.5 are precisely—and the only ones—fit for use in energy formulas featuring atomic charges.

The accuracy of the results obtained from eq. (6.10) is best illustrated by comparisons between calculated and experimental atomization energies. The latter are deduced from experimental enthalpies of formation and zero-point plus heat-content energies, as indicated in Appendix A, eq. (A.6). These results are self-explanatory (Table 6.5). They certainly support the basic energy theory and validate the quality of the charges used in its applications.

The accuracy demonstrated in Table 6.5 renders this bond-energy theory attractive for practical applications, but there is a drawback: the time-consuming evaluation of the nonbonded part, E_{nb}. There is a simple way of by-passing the explicit calculation of E_{nb}, meaning that we need not any longer compute molecular geometries and the charges of the individual hydrogen atoms.

Consider eq. (6.10) and rewrite it as follows:

$$\Delta E_a^* = (1-m)(\varepsilon_{CC}^\circ + 6\varepsilon_{CH}^\circ)$$
$$+(n-2+2m)(\varepsilon_{CC}^\circ + 6\varepsilon_{CH}^\circ - 4\varepsilon_{CH}^\circ + a_{HC}\, q_H^\circ - A_2 \delta_C^{CH_4})$$
$$+(n-2+2m)A_2\delta_C^{CH_4} + A_1 \sum N_{CC}\delta_C + A_2 \sum \delta_C - E_{nb}.$$

Using the quantities $\Delta E_{a,bd}^{*C_2H_6}$ and $\Delta E_{a,bd}^{*CH_4}$ defined above, eq. (6.10) becomes

$$\Delta E_a^* = (1-m)\Delta E_{a,bd}^{*C_2H_6} + (n-2+2m)\left(\Delta E_{a,bd}^{*C_2H_6} - \Delta E_{a,bd}^{*CH_4}\right)$$
$$(n-2+2m)A_2\delta_C^{CH_4} + A_1 \sum N_{CC}\delta_C + A_2 \sum \delta_C - E_{nb}.$$

Finally, using the approximation (6.9) for E_{nb} and applying (3.5) we have

$$\Delta E_a^* \simeq (1-m)\Delta E_a^{*C_2H_6} + (n-2+2m)\left(\Delta E_a^{*C_2H_6} - \Delta E_a^{*CH_4}\right)$$
$$(n-2+2m)A_2\delta_C^{CH_4} + A_1 \sum N_{CC}\delta_C + A_2 \sum \delta_C. \qquad (6.12)$$

Now we know that nonbonded energies differ somewhat from case to case in comparisons between structural isomers. On the other hand, the $\sum N_{CC}\delta_C$ and $\sum \delta_C$ terms are also structure-dependent. For these reasons, a minor readjustment of the parameters A_1 and A_2 succeeds in compensating part of the error introduced by assuming exact additivity for the nonbonded Coulomb contributions. The recommended values are [20], in kcal mol^{-1} ppm^{-1} units, $A_1 = 0.03244$ and $A_2 = 0.05728$. With $\Delta E_a^{*C_2H_6} = 710.54$ and $\Delta E_a^{*CH_4} = 419.27$ kcal mol^{-1}, equation (6.12) becomes, in kcal mol^{-1} units,

$$\Delta E_a^* \simeq 710.54(1-m) + 290.812(n-2+2m)$$
$$+0.03244 \sum N_{CC}\delta_C + 0.05728 \sum \delta_C. \qquad (6.13)$$

This handy energy formula only requires the ^{13}C NMR spectrum of the molecule under consideration. The standard enthalpies of formation are deduced from eq. (A.6), using the zero-point plus heat-content energies given by (A.9) for the open-chain alkanes, or by (A.16) for the cycloalkanes. The results[7] of Table 6.6[8] certainly justify the use of approximate nonbonded energies leading to equation (6.13).

---

[7]Many additional results are reported in [3].

[8]The sources of the NMR input data and of the experimental enthalpies of formation are reported in [3, 20]. The NMR shifts of diamantane are taken from [21] and its ΔH_f° from [22]. The NMR results for trans- and cis-1,2-dimethylcyclohexane are taken from [23] and their experimental ΔH_f° values are from [24].

TABLE 6.6. Comparison between calculated and experimental enthalpies of formation, ΔH_f^o(298.15, gas), of saturated hydrocarbons, kcal mol^{-1}

Molecule	$\sum N_{CC}\delta$	$\sum \delta$	ΔE_a^*	ΔH_f^o Calcd.	ΔH_f^o Exptl.
Methane	0	-8	419.27	-17.73	-17.89
Ethane	0	0	710.54	-20.15	-20.04
Propane	39.8	29.6	1004.34	-25.11	-25.02
Butane	91.0	52.8	1298.14	-30.07	-30.03
Isobutane	113.4	74.8	1300.13	-32.40	-32.07
Pentane	139.6	77.6	1591.95	-35.05	-35.00
Isopentane	161.6	87.7	1593.24	-36.68	-36.92
Hexane	188.4	102.2	1885.75	-39.98	-39.96
2-Methylpentane	208.9	114.2	1887.11	-41.72	-41.66
3-Methylpentane	209.5	101.2	1886.38	-40.99	-41.02
2,3-Dimethylbutane	223.8	110.6	1887.38	-42.33	-42.49
Heptane	233.0	124.3	2179.28	-44.85	-44.89
2-Methylhexane	257.5	138.2	2180.87	-46.63	-46.60
3-Methylhexane	255.5	126.7	2180.15	-45.91	-45.96
2,4-Dimethylpentane	274.6	151.3	2182.17	-48.28	-48.30
Nonane	332.0	174.1	2766.97	-54.71	-54.54
4-Methyloctane	350.7	177.3	2767.76	-55.85	-56.19
2,3,5-Trimethylhexane	378.0	185.4	2769.11	-57.88	-57.97
Cyclohexane	261.6	130.8	1760.85	-29.36	-29.50
Methylcyclohexane	349.8	169.7	2056.75	-36.71	-36.98
Ethylcyclohexane	402.1	186.8	2350.26	-41.33	-41.05
n-Butylcyclohexane	492.1	234.3	2937.52	-50.83	-50.92
trans-1,2-Dimethylcyclohexane	440.0	200.7	2352.27	-43.68	-43.00
cis-1,2-Dimethylcyclohexane	369.5	166.1	2348.00	-41.12	-41.13
cis-1,3-Dimethylcyclohexane	437.0	208.5	2352.63	-44.02	-44.13
trans-1,3-Dimethylcyclohexane	373.9	180.4	2348.97	-42.06	-42.18
trans-1,4-Dimethylcyclohexane	436.8	208.5	2352.62	-44.01	-44.10
cis-1,4-Dimethylcyclohexane	378.7	179.3	2349.07	-42.16	-42.20
1,1-Dimethylcyclohexane	391.4	194.5	2350.35	-43.12	-43.23
1-cis-3-cis-5-Trimethylcyclohexane	527.3	248.9	2648.69	-51.52	-51.48
1-cis-3-trans-5-Trimethylcyclohexane	462.6	223.1	2645.11	-49.65	-49.37
trans-Decalin	630.8	277.1	2815.56	-43.74	-43.52
cis-Decalin	527.0	232.5	2809.64	-40.38	-40.43
trans-syn-trans-Perhydroanthracene	997.4	422.9	3870.15	-58.01	-58.13
trans-anti-trans-Perhydroanthracene	900.5	389.2	3865.08	-52.94	-52.74
Bicyclo[2.2.2]octane	363.7	163.1	2218.74	-24.09	-23.75
Adamantane	662.3	285.4	2688.12	-30.34	-30.65
Diamantane	1095.6	431.0	3615.93	-32.49	-32.60
Twistane	514.4	211.8	2679.14	-21.36	-21.60
Spiro[5.5]undecane	552.2	249,4	3102.24	-44.93	-44.81

6.6   Unsaturated hydrocarbons

The merits of the theory are adequately tested with the help of eq. (3.36)

$$\sum_{k<l} \varepsilon_{kl} = \sum_{k<l} \varepsilon_{kl}^{\circ} + \sum_{k} \sum_{l} a_{kl} \Delta q_k + F$$

and the approximation $\Delta E_a^* \simeq \sum_{k<l} \varepsilon_{kl}$. Explicit evaluations of the non-bonded terms are not contemplated because the straightforward use of (3.11) cannot be justified for unsaturated hydrocarbons[9]. The numerical verifications are thus not as exhaustive as desired, but this circumstance does not obscure the general validity of the leading terms, $\sum_k \sum_l a_{kl} \Delta q_k$ and F, which are calculated from theory.

Two approaches are possible. They are compactly summarized in eq. (3.37). In a nutshell, for simple olefins we use the left-hand side of (3.37) because it offers a simpler overview of the applicability of general theory. For dienes and aromatic hydrocarbons, in contrast, it is easier to use the right-hand side of (3.37), i.e., appropriately modified reference bond energies, because (3.36) then reduces to the simpler form (3.38). Let us see how it works.

Olefins, C_nH_{2n}

The $\sum \varepsilon_{kl}^{\circ}$ part is constructed from ε_1°, ε_{10} (which are the basic CC and CH reference bond energies of ethane) and ε_2°, for the ethylenic double bond. Now we seek a general expression for $\sum_k \sum_l a_{kl} \Delta q_k + F$.

The first thing to do is to define the charges entering these calculations. Let $q_C^{\circ C_2 H_6} = 35.1$ me represent the reference net charge of the ethane carbon and $q_C^{\circ C_2 H_4} = 7.7$ me that of ethylene. Their difference is

$$\Delta q_C^{\circ} = q_C^{\circ C_2 H_4} - q_C^{\circ C_2 H_6} = -27.4 \text{ me}. \qquad (6.14)$$

The sp^3 carbon atoms are simply represented by C. The sp^2 carbons are identified by the subscript "α". So we write:

$$\Delta q_C = q_C - q_C^{\circ C_2 H_6} \quad \text{for } sp^3 \text{ C atoms} \qquad (6.15)$$

$$\Delta q_{C_\alpha} = q_{C_\alpha} - q_C^{\circ C_2 H_4} \quad \text{for } sp^2 \text{ C atoms.} \qquad (6.16)$$

---

[9]Approximate corrections could be envisaged for the saturated portion, but it is felt that this would not add to the overall comprehension of the results, in spite of numerical improvements between calculated and experimental quantities.

Finally, the difference

$$q_{C_\alpha} - q_C^{oC_2H_6} = q_{C_\alpha} - q_C^{oC_2H_4} + q_C^{oC_2H_4} - q_C^{oC_2H_6}$$

gives

$$q_{C_\alpha} - q_C^{oC_2H_6} = \Delta q_{C_\alpha} + \Delta q_C^o \qquad (6.17)$$

where Δq_C^o is the charge difference defined by (6.14).

For the hydrogen atoms, of course, $\Delta q_H = q_H - q_H^o$, where $q_H^o = -11.7$ me is the hydrogen net charge of ethane, selected as reference for *all* the H atoms. Their number is $2n$, with $n =$ number of the C atoms in the molecule. Charge normalization, $\sum q_H = -\sum q_C - \sum q_{C_\alpha}$, and eqs. (6.14)–(6.17) give

$$\sum \Delta q_H = -\left(\sum \Delta q_C + \sum \Delta q_{C_\alpha}\right) - 2\Delta q_C^o + nq_H^o. \qquad (6.18)$$

Equation (6.18) conveniently eliminates explicit calculations of hydrogen charges from the general expression for $\sum_k \sum_l a_{kl}\Delta q_k + F$.

In the following, N_{CC} is the number of CC bonds and N_{CH} is the number of CH bonds formed by an sp^3 carbon atom. $N_{C_\alpha C}$ and $N_{C_\alpha H}$ are the numbers of CC and CH bonds, respectively, formed by sp^2 carbon atoms. Note that $N_{CC} + N_{CH} = 4$ and that $N_{C_\alpha C} + N_{C_\alpha H} = 2$. Now we use (3.36). For the carbon atoms we have

$$\overset{\text{CC bonds}}{\sum_k \sum_l} a_{kl}\Delta q_k = a_{CC} \sum N_{CC}\Delta q_C + a_{C=C} \sum \Delta q_{C_\alpha}$$
$$+ a_{C_\alpha C} \sum N_{C_\alpha C}\left(q_{C_\alpha} - q_C^{oC_2H_6}\right). \qquad (6.19)$$

Similarly, we obtain for the CH bonds that

$$\overset{\text{CH bonds}}{\sum_k \sum_l} a_{kl}\Delta q_k = a_{CH} \sum N_{CH}\Delta q_C + a_{C_\alpha H} \sum N_{C_\alpha H}\left(q_{C_\alpha} - q_C^{oC_2H_6}\right)$$
$$+ a_{HC} \sum \Delta q_H. \qquad (6.20)$$

For F we use F_{CC} (described as F in example 1) and F_{CH} (described in example 2) and obtain the simple result

$$F = F_{CC} \sum N_{C_\alpha C} + F_{CH} \sum N_{C_\alpha H}. \qquad (6.21)$$

The final energy formula follows from eqs. (6.16)–(6.21) and is:

TABLE 6.7. Verification of eq. (6.22) for *trans*- and *cis*-olefins, kcal mol^{-1}

Parameter	*trans*-Olefins		*cis*-Olefins	
	Theor.	Empir.	Theor.	Empir.
$4F_{CH} + (4a_{C_\alpha H} - 2a_{HC})\Delta q_C^o$	-19.08	-19.16	-18.27	-18.22
$(a_{C_\alpha C} - a_{C_\alpha H})\Delta q_C^o + F_{CC} - F_{CH}$	4.15	4.19	3.95	4.0

$$\sum_k \sum_l a_{kl}\Delta q_k + F = A_1 \sum N_{CC}\Delta q_C + A_2 \sum \Delta q_C + A_{1\alpha} \sum N_{C_\alpha C}\Delta q_{C_\alpha}$$

$$+ A_3 \sum \Delta q_{C_\alpha} + \left[4F_{CH} + (4a_{C_\alpha H} - 2a_{HC})\Delta q_C^o\right]$$

$$+ \left[(a_{C_\alpha C} - a_{C_\alpha H})\Delta q_C^o + F_{CC} - F_{CH}\right]\sum N_{C_\alpha C}$$

$$+ a_{HC}\, n q_H^o \qquad\qquad (6.22)$$

where

$$A_1 = a_{CC} - a_{CH}$$
$$A_2 = 4a_{CH} - a_{HC}$$
$$A_{1\alpha} = a_{C_\alpha C} - a_{C_\alpha H}$$
$$A_3 = a_{C=C} + 2a_{C_\alpha H} - a_{HC}.$$

Equation (6.22) lends itself to numerical verifications. The first two terms are well known: they are those described for the saturated hydrocarbons and are calculated in precisely the same fashion, with $A_1 = 0.0356$ and $A_2 = 0.0529$ kcal mol^{-1} ppm^{-1}. The last term, with $a_{HC} = -1.007$ au, is $7.393n$ kcal mol^{-1}. Using this theoretical input and the appropriate sums $\sum \varepsilon_{kl}^o$ in comparisons with experimental atomization energies, one obtains $A_{1\alpha}$ and A_3 and the empirical estimates of the two terms in brackets reported in Table 6.7. The latter can be evaluated theoretically using the F_{CC} and F_{CH} results given in examples 1 and 2, respectively, and the appropriate $a_{C_\alpha C}(1.531)$ and $a_{C_\alpha H}(1.08)$ parameters described in examples 6 and 12, respectively, with $\Delta q_C^o = -27.4$ me, eq. (6.14). The empirical results nicely confirm the theoretical predictions. Note that the latter cover the largest part, by far, of all the energy terms occurring in ΔE_a^*. The unresolved part of (6.22), i.e., $A_{1\alpha}\sum N_{C_\alpha C}\Delta q_{C_\alpha} + A_3\sum \Delta q_{C_\alpha}$, represents less than 1 kcal mol^{-1} and can be made responsible only for a small blur in the present assessment regarding the validity of (6.22).

The calculation of these unresolved terms involves the formulas given in Chapter 3.7. Equation (3.59) tells us that

$$A_{1\alpha} = \frac{mA_{1\alpha}^{\sigma} + A_{1\alpha}^{\pi}}{m+1} \quad \text{and} \quad A_3 = \frac{mA_3^{\sigma} + A_3^{\pi}}{m+1}.$$

On the other hand, using the relationship $\Delta q_{C_\alpha} = (m+1)/(d\delta_{C_\alpha}/dq^\pi)\delta_{C_\alpha}$ between NMR shifts, δ_{C_α} ppm from ethene, and Δq_{C_α} (Chapter 5.6), we get

$$A_{1\alpha}\Delta q_{C_\alpha} = \frac{mA_{1\alpha}^{\sigma} + A_{1\alpha}^{\pi}}{(d\delta_{C_\alpha}/dq^\pi)}\delta_{C_\alpha} \tag{6.23}$$

$$A_3\Delta q_{C_\alpha} = \frac{mA_3^{\sigma} + A_3^{\pi}}{(d\delta_{C_\alpha}/dq^\pi)}\delta_{C_\alpha}. \tag{6.24}$$

The limited precision of present charge analyses of olefins does not permit reliable direct evaluations of m and $(d\delta_{C_\alpha}/dq^\pi)$. The empirical regression using experimental atomization energies and all the known theoretical parameters of (6.22) indicates that $A_{1\alpha}\Delta q_{C_\alpha} \simeq -0.028\delta_{C_\alpha}$ and $A_3\Delta q_{C_\alpha} \simeq 0.20\delta_{C_\alpha}$ kcal mol^{-1}, suggesting that $m \simeq -0.955$, $(d\delta_{C_\alpha}/dq^\pi) \simeq 300$ ppm/e and $\Delta q_{C_\alpha} \simeq 0.15\delta_{C_\alpha}$ me [25]. We can use these estimates, *faute de mieux*, but should not attach too much importance to this type of numerical results. It is fortunate, indeed, that the contribution of this empirical part of (6.22) turns out to be very small. Under these circumstances, we may as well temporarily use a simplifying substitute for $A_{1\alpha}\sum N_{C_\alpha C}\Delta q_{C_\alpha} + A_3\sum \Delta q_{C_\alpha}$ by taking $0.18\sum \delta_{C_\alpha}$ to cover this portion of (6.22) [25]. The approximations

$$\sum_k\sum_l a_{kl}\Delta q_k \simeq 0.0356\sum N_{CC}\delta_C + 0.0529\sum \delta_C + 0.18\sum \delta_{C_\alpha}$$
$$+7.393n + 4.19\sum N_{C_\alpha C} - 19.16 \text{ kcal mol}^{-1} \tag{6.25}$$

for ethylene, 1-alkenes, *trans*-alkenes and tetramethylethylene, and

$$\sum_k\sum_l a_{kl}\Delta q_k \simeq 0.0356\sum N_{CC}\delta_C + 0.0529\sum \delta_C + 0.18\sum \delta_{C_\alpha}$$
$$+7.393n + 4.0\sum N_{C_\alpha C} - 18.22 \text{ kcal mol}^{-1} \tag{6.26}$$

for *cis*-olefins appear to be quite adequate in practical applications (Table 6.8). The enthalpies are obtained as indicated in Appendix A, with the help of the zero-point plus heat-content energies described therein.

TABLE 6.8. Comparison between calculated and experimental enthalpies of formation, ΔH_f^o(298.15, gas), of olefinic hydrocarbons, kcal mol^{-1}

Molecule	$\sum_{k<l} \varepsilon_{kl}^o$	$\sum_k \sum_l a_{kl} \Delta q_k$	ΔE_a^*	ΔH_f^o Calcd.	ΔH_f^o Exptl.
Ethene	566.59	−4.37	562.22	12.38	12.50
Propene	849.84	8.80	858.64	4.80	4.88
1-Butene	1133.08	19.47	1152.55	−0.27	−0.03
(Z)2-Butene	1133.08	20.74	1153.83	−1.54	−1.67
(E)2-Butene	1133.08	21.91	1154.99	−2.71	−2.67
2-Me-Propene	1133.08	23.42	1156.51	−4.22	−4.04
1-Pentene	1416.33	30.08	1446.41	−5.28	−5.00
(Z)2-Pentene	1416.33	31.55	1447.88	−6.75	−6.71
(E)2-Pentene	1416.33	32.39	1448.72	−7.60	−7.59
2-Me-1-Butene	1416.33	33.84	1450.17	−9.05	−8.68
3-Me-1-Butene	1416.33	31.33	1447.66	−6.87	−6.92
2-Me-2-Butene	1416.33	34.98	1451.31	−10.19	−10.17
1-Hexene	1699.57	40.18	1739.76	−9.80	−9.96
(Z)2-Hexene	1699.57	42.02	1741.60	−11.64	−12.51
(E)2-Hexene	1699.57	42.99	1742.57	−12.61	−12.88
(Z)3-Hexene	1699.57	42.28	1741.86	−11.90	−11.38
(E)3-Hexene	1699.57	43.00	1742.58	−12.62	−13.01
2-Me-1-Pentene	1699.57	44.25	1743.82	−13.86	−14.19
3-Me-1-Pentene	1699.57	41.65	1741.22	−11.60	−11.82
4-Me-1-Pentene	1699.57	41.94	1741.52	−11.90	−12.24
(Z)3-Me-2-Pentene	1699.57	45.71	1745.29	−15.33	−15.08
(E)3-Me-2-Pentene	1699.57	45.14	1744.71	−14.75	−14.86
(Z)4-Me-2-Pentene	1699.57	43.59	1743.16	−13.54	−13.73
(E)4-Me-2-Pentene	1699.57	44.43	1744.00	−14.38	−14.69
2-Et-1-Butene	1699.57	43.34	1742.91	−12.95	−13.38
2,3-diMe-1-Butene	1699.57	45.37	1744.95	−15.33	−15.85
3,3-diMe-1-Butene	1699.57	43.31	1742.89	−13.61	−14.70
2,3-diMe-2-Butene	1699.57	47.17	1746.74	−16.78	−16.68
(Z)2-Heptene	1982.82	52.47	2035.29	−16.49	−16.9
(E)2-Heptene	1982.82	53.43	2036.25	−17.44	−17.6
(Z)3-Heptene	1982.82	52.76	2035.58	−16.77	−16.90
(E)3-Heptene	1982.82	53.59	2036.40	−17.60	−17.60
(Z)3-Me-3-Hexene	1982.82	55.85	2038.66	−19.86	−18.60
(E)3-Me-3-Hexene	1982.82	55.92	2038.74	−19.93	−19.22
2,4-diMe-1-Pentene	1982.82	56.38	2039.19	−20.74	−20.27
4,4-diMe-1-Pentene	1982.82	54.65	2037.47	−19.35	−19.20
(E)4,4-diMe-2-Pentene	1982.82	56.98	2039.70	−21.58	−21.46
(E)2,2-diMe-3-Hexene	2266.07	67.45	2333.52	−26.56	−26.16
2-Me-3-Et-1-Pentene	2266.07	66.06	2332.12	−24.82	−24.40

The comparison between theory and experiment[10] certainly permits to appreciate the former at its merits. In particular, the idea of using approximate *transferable* F_{CC} and F_{CH} bond terms for general use seems justified by the present results.

Dienes

The general idea is simple. It exploits eq. (3.38). The sum $\sum \varepsilon_{kl}^0$ is calculated using ε_1^0 and ε_{10}^0 for the $C(sp^3)$-$C(sp^3)$ and $C(sp^3)$-H bonds, respectively, and ε_2^0 for the double bonds. For the $C(sp^2)$-$C(sp^2)$ single bond, one takes ε_5^0 and, finally, uses ε_4^0 and ε_{11}^0 for the $C(sp^3)$-$C(sp^2)$ and $C(sp^2)$-H bonds, respectively. No separate calculation of F is required, for it is included in the modified references ε_4^0, ε_5^0 and ε_{11}^0 following the protocol described in eq. (3.57). In this manner, (3.36) reduces to the simpler form (3.38).

The calculation of $\sum_k \sum_l a_{kl} \Delta q_k$ is equally simple. It involves the following steps. *i*) First we calculate $\sum_l a_{kl}$ for each carbon atom and the appropriate Δq_C . So we obtain $\Delta q_C \times \sum_l a_{kl} = \sum_l a_{kl} \Delta q_k$ for each individual C_k of the molecule. The sum of all these $\sum_l a_{kl} \Delta q_k$ terms gives the total contribution of all the C atoms to the final sum $\sum_k \sum_l a_{kl} \Delta q_k$. *ii*) Using the same Δq_C 's, one forms their sum $\sum \Delta q_C$. Knowing the reference charges, 35.1 me for the sp^3 carbons and 7.7 me for the sp^2 C atoms, one deduces the total net charge, $\sum q_C$, of all the carbons. *iii*) Thus we have $\sum q_H = -\sum q_C$ and, hence, $\sum \Delta q_H = \sum q_H - n_H q_H^0$, n_H being the number of H atoms. So we get $a_{HC} \sum \Delta q_H$ which is the total contribution of all the hydrogen atoms to $\sum_k \sum_l a_{kl} \Delta q_k$. *iv*) The final sum $\sum_k \sum_l a_{kl} \Delta q_k$ is obtained by adding the contributions of the carbon and the hydrogen atoms.

This approach is probably the most convenient one for general use. It could have been easily applied to the saturated and ethylenic hydrocarbons examined earlier: as a rule, one can always use the atom-by-atom method described here, with modifications, if necessary, depending on the class of molecules investigated, the limiting factor being our knowledge about the proper atomic charges. Here we use this atom-by-atom construction of $\sum_k \sum_l a_{kl} \Delta q_k$ with a confidence build upon the experience gathered with the $C_nH_{2n+2-2m}$ and C_nH_{2n} molecules. The results are collected in Table 6.9[11]. They are self-explanatory.

---

[10]Note that the $\sum_k \sum_l a_{kl} \Delta q_k$ results of Table 6.8 *include* F. The experimental ΔH_f^0 values are taken from Ref.[26]. The NMR data are from [27]-[29]. The assignments of the Z and E isomers of 3-methyl-3-hexene and of 3-methyl-2-pentene are those given in [30].

[11]The sources of the ΔH_f^0 values and the appropriate ^{13}C NMR shifts are given in [31].

TABLE 6.9. Comparison between calculated and experimental atomization energies of selected dienes, kcal mol^{-1}

Molecule	ΔH_f^o	$\sum_{k<l} \varepsilon_{kl}^o$	$\sum_k \sum_l a_{kl}\Delta q_k$	ΔE_a^* Calcd.	ΔE_a^* Exptl.
1,3-Butadiene	26.33	1032.02	−22.41	1009.6	1010.1
(Z)1,3-Pentadiene	19.77	1319.42	−14.09	1305.3	1305.4
(E)1,3-Pentadiene	18.77	1319.42	−12.85	1306.6	1306.5
Isoprene	18.06	1319.42	−13.09	1306.3	1306.7
1,4-Pentadiene	25.25	1311.83	−11.61	1300.2	1300.1
1,5-Hexadiene	20.11	1595.08	−0.99	1594.1	1594.0
Dimethyl-1,3-butadiene	10.78	1606.82	−4.20	1602.6	1602.8
1,3-Cyclohexadiene	25.38	1462.84	10.98	1473.8	1473.7

Benzenoid hydrocarbons

The calculations are made following the strategy explained for the dienes.

The $\sum \varepsilon_{kl}^o$ part is constructed using the appropriate reference bond energies of Table 6.3. A comment is in order as regards ε_8^o, for a bond between aryl carbons at a distance of 1.397 Å. Consider the aromatic CC bond, ε_3^o. It represents, so to speak, the average between a single and a double bond as they are found in benzenoid structures: it is the CC bond of benzene. It is counted twice the number of double bonds one can write using classical Kékulé structures, e.g., 10 times for naphthalene, 14 times for anthracene, etc. The remaining CC bonds (e.g., 1 in naphthalene, 2 in anthracene, etc.) are treated as $C(sp^2)$–$C(sp^2)$ single bonds: these are the bonds decribed by ε_8^o. Of course, no bond in particular is identified in this manner, it is only the number of bonds that matters. Incidentally, note that the picture describing a CC bond of benzene as something resembling the average between a single and a double bond is roughly satisfied, provided one uses the sp^2–sp^2 single bond, ε_8^o, in this average and not that of ethane, as has been done in the past: the average of ε_2^o and ε_8^o, 115.29 kcal mol^{-1}, vividly supports this interpretation. But it is equally clear that one cannot simply use ε_3^o for each CC bond found in aromatic cycles because there are not as many 'averages' as there are CC bonds, except in benzene itself.

Three parameters need be introduced. i) The reference bond energy ε_3^o describing the CC bonds of benzene. This is the only empirically adjusted bond-energy term, all the other bonds—and there is quite a variety of them—being calculated as indicated in Section 3. SCF analyses like those presented in example 5 consistently show that this bond energy is $\sim$1.65 times that of ethane [16][12], thus supporting the value adopted here, 115.39 kcal mol^{-1}.

ii) The relationship $\Delta q_C \simeq 1.2\delta_C$ me between aryl-C charges and ^{13}C NMR shifts, where both Δq_C and δ_C are expressed with reference to the benzene C atom. This relationship, suggesting $m = -0.8137$ for $d\delta/dq^\pi = 157$ ppm/e, eq. (3.59), holds for aromatic carbons bonded to H and to conjugated C atoms and was obtained from SCF charge analyses [12]: this result is admittedly crude, but extensive numerical analyses gave no reason for revision[13].

iii) A different value of m, namely, $m = -0.8947$, for any aryl-C atom bonded to a nonconjugated moiety (like the methyl carbon of toluene or in the central bond of biphenyl). The formula $\Delta q_C = (m+1)/(d\delta_C/dq^\pi)\delta_C$ now gives $\Delta q_C \simeq 0.67\delta_C$, assuming that $d\delta_C/dq^\pi = 157$ me is still valid. This m value is firmly assessed by numerical applications to alkyl-substituted benzenes, but its *a priori* evaluation resists presently available charge analyses.

The a_{Cl}^σ and a_{Cl}^π parameters are calculated in the usual fashion with the help of eq. (6.6) and Table 6.2.

Example 13. Typical values are $a_{CC}^\sigma(1.40) = -0.485_0$ and $a_{CC}^\pi(1.40) = -0.461_4$ kcal mol^{-1} me^{-1} for the aromatic CC bonds, and $a_{CH}^\sigma(1.08) = -0.210_2$ and $a_{CH}^\pi(1.08) = -0.186_6$ kcal mol^{-1} me^{-1} for the CH bonds. Using eq. (3.59) with $m = -0.8137$ one finds $a_{CC}^{\sigma\pi} = -0.358_3$ and $a_{CH}^{\sigma\pi} = -0.083_5$ kcal mol^{-1} me^{-1} for these bonds. On the other hand, for the $C(Ar)-C(sp^3)$ bond like that occurring in toluene, we have for the aryl carbon $a_{CC}^\sigma(1.48) = -0.463_1$ and $a_{CC}^\pi(1.48) = -0.439_4$ kcal mol^{-1} me^{-1}. Hence it is now $a_{CC}^{\sigma\pi} = -0.238_5$ kcal mol^{-1} me^{-1}.

The atom-by-atom calculation of $\sum_k \sum_l a_{kl}\Delta q_k$ is made following the recipe developed for the dienes. The final results are given in Table 6.10[14].

Attention must be given to the planarity (or lack of it) of the benzenoid structures. 4,5-Dimethylphenanthrene, for example, is evaluated at $\Delta H_f^\circ = 36.8$ kcal mol^{-1} assuming planarity. Neighboring methyl groups which are separated by five bonds in the molecular skeleton provoke chiral nonplanar conformations [33]. Modeling, where appropriate, the CC bonds on those describing biphenyl and *cis*-stilbene, one predicts [12] $\Delta H_f^\circ = 47.8$ for the nonplanar form, in acceptable agreement with the reported value [24] 46.26$\pm$1.46 kcal mol^{-1}. Additional examples concerning, e.g., the nonplanar triphenylene [34]) and the dimethylnaphthalene isomers, are offered in [12].

---

[12]The results obtained using Pople's 6-311G** and 6-311G($2df, 2pd$) bases [14] and Dunning's cc-pVTZ basis [32] are virtually the same.

[13]Empirical fittings using experimental atomization energies and the energy formula (3.38) indicate a factor of 1.19 instead of 1.2.

[14]The sources of the experimental ΔH_f°'s and of the NMR shifts used in these calculations are reported in [12].

TABLE 6.10. Comparison between calculated and experimental atomization
energies of selected aromatic hydrocarbons, kcal mol^{-1}

Molecule	ΔH_f°	$\sum \epsilon_{kl}^\circ$	$\sum_{k,l} a_{kl} \Delta q_k$	ΔE_a^* Calcd.	ΔE_a^* Exptl.[a]
Benzene	19.81	1360.80	5.69	1366.5	1366.5
Toluene	11.99	1649.14	14.17	1663.3	1663.2
1,2-Dimethylbenzene	4.56	1937.48	22.23	1959.7	1959.4
1,3-Dimethylbenzene	4.14	1937.48	22.51	1960.0	1959.9
1,4-Dimethylbenzene	4.31	1937.48	21.91	1959.4	1959.7
1,2,3-Trimethylbenzene	−2.26	2225.81	29.70	2255.5	2255.1
1,2,4-Trimethylbenzene	−3.31	2225.81	29.92	2255.7	2256.2
1,3,5-Trimethylbenzene	−3.81	2225.81	30.57	2256.4	2256.6
1,2,3,4-Tetramethylbenzene	−10.02	2514.15	38.50	2552.6	2551.7
1,2,3,5-Tetramethylbenzene	−10.71	2514.15	38.03	2552.2	2552.4
1,2,4,5-Tetramethylbenzene	−10.82	2514.15	37.41	2551.6	2552.5
Pentamethylbenzene	−17.80	2802.49	45.78	2848.3	2848.3
Hexamethylbenzene	−25.26	3090.83	54.30	3145.1	3144.6
Ethylbenzene	7.15	1932.38	24.03	1956.4	1956.9
n-Propylbenzene	1.89	2215.63	34.71	2250.3	2250.9
Isopropylbenzene	0.96	2215.63	36.12	2251.8	2251.5
sec-Butylbenzene	−4.15	2498.87	46.17	2545.0	2545.5
tert-Butylbenzene	−5.40	2498.87	48.32	2547.2	2546.7
1,2-Diphenylethane	32.4	3154.30	48.27	3202.6	3202.9
Styrene	35.30	1810.42	0.58	1811.0	1810.7
cis-Stilbene	60.31	3034.23	22.60	3056.8	3056.8
trans-Stilbene	52.5	3039.83	23.53	3063.4	3064.6
Biphenyl	43.53	2587.81	24.86	2612.7	2612.4
Naphthalene	36.25	2136.39	21.19	2157.6	2157.6
1-Methylnaphthalene	27.93	2424.73	30.60	2455.3	2454.8
2-Methylnaphthalene	27.75	2424.73	29.84	2454.6	2454.9
2,6-Dimethylnaphthalene	See text	2713.07	38.25	2751.3	2751.0
Anthracene	55.2	2911.98	34.46	2946.4	2946.3
Phenanthrene	49.5	2911.98	40.66	2952.6	2952.0
Pyrene	53.94	3233.97	61.64	3295.6	3295.7
Triphenylene	61.9	3681.03	66.83	3747.9	3747.1
Benz[a]anthracene	65.97	3687.57	55.46	3743.0	3743.0
7,12-Dimethylbenz[a]anthracene	66.4	4264.25	71.63	4335.9	4320.2
Dibenz[a,c]anthracene		4463.16	76.17	4539.3	(4540.5)
Dibenz[a,h]anthracene		4463.16	73.72	4536.9	(4538.0)
1,2,3,4-Tetrahydronaphthalene	7.3	2360.00	60.15	2420.2	2420.2
9,10-Dihydroanthracene	38.2	3020.50	62.17	3082.7	3083.1

[a]The atomization energies indicated in parentheses are theoretical results deduced from
enthalpies of formation calculated by Dewar and de Llano [35]. The sources of the
experimental enthalpies of formation are indicated in [12]. Additional results are given
in reference [12].

In closing, let us examine graphite. It has a layer-like structure. Each carbon is bonded to three other carbon atoms forming a framework of planar benzenoid rings, with bond distances of 1.42 Å [36]. Two of these bonds command the use of ε_3^o while ε_8^o must be used for the third one. Now we calculate $a_{CC}\Delta q_C$. The graphite carbons are electroneutral. Hence $\Delta q_C = -13.2$ me. The bond lengthening from 1.40 to 1.42 Å modifies $a_{CC}^{\sigma\pi}$ from -0.358_3 to -0.352_6 kcal mol^{-1} me^{-1}. The compromise adopted here, -0.355_4, gives $2 \times a_{CC}^{\sigma\pi}\Delta q_C = 9.38$ kcal mol^{-1} and thus

$$\Delta E_a^*(\text{graphite}) = \frac{1}{2}(2\varepsilon_3^o + \varepsilon_8^o + 3 \times 9.38) = 175.07 \quad \text{kcal mol}^{-1}.$$

The standard enthalpy of atomization

$$\Delta H_a^o(\text{graphite}, 298.15\text{K}) = 171.31 \quad \text{kcal mol}^{-1}$$

is deduced from (A.6) using the calculated zero-point and the recommended heat-content energies, 3.68 [37] and 1.56 [38] kcal mol^{-1}, respectively. Our calculation neglects the inteactions between the layers, of the order of ~1.2 kcal mol^{-1} [39], but is surely in acceptable agreement with the experimental value, $\Delta H_a^o(298.15\text{K}) = 171.29 \pm 0.11$ kcal mol^{-1} [38, 40].

6.7   Ethers and Carbonyl compounds

Two new reference bond energies are required at this point: *i)* ε_{13}^o for the CO single bond of diethylether and *ii)* ε_{14}^o for the CO double bond of acetone, namely,

$$\varepsilon_{13}^o = 79.78 \text{ kcal mol}^{-1} \text{ for } q_{C_\alpha}^o = 31.2_6 \text{ me and } q_O^o = 5.1_8 \text{ me}$$
$$\varepsilon_{14}^o = 179.40 \text{ kcal mol}^{-1} \text{ for } q_{C_\alpha}^o = 14.0 \text{ me and } q_O^o = -21.2 \text{ me}.$$

The subscript "α" identifies the carbons adjacent to oxygen. These values yield the best fits with experimental energies [3, 41]. Direct estimates using eqs. (3.16) and (3.18) and the appropriate SCF potentials at the nuclei suggest $\varepsilon_{13}^o \simeq 80$ and $\varepsilon_{14}^o \simeq 181$ kcal mol^{-1} [41][15]. (The charges are described in Chapter 5.7.) The CC and CH bonds are treated in the usual manner, with reference to ε_{CC}^o and ε_{CH}^o, respectively.

---

[15]The agreement between the two sets of data is certainly encouraging, except for the unfortunate fact that theoretical evaluations of this sort are clouded by approximations and should be taken *cum grano salis*.

The relevant charge variations are $\Delta q_C = q_C - q_C^{\circ C_2 H_6}$ for C atoms not bonded to oxygen, $\Delta q_{C_\alpha} = q_{C_\alpha} - q_{C_\alpha}^\circ$ for those adjacent to O, and $\Delta q_O = q_O - q_O^\circ$ for oxygen. For future use, let us define $\Delta q_C^\circ = q_{C_\alpha}^\circ - q_C^{\circ C_2 H_6}$:

$$\Delta q_C^\circ \simeq -3.8_4 \text{ me for diethylether}$$
$$\Delta q_C^\circ \simeq -21.1 \text{ me for acetone.}$$

Also note that $q_{C_\alpha} - q_C^{\circ C_2 H_6} = \Delta q_{C_\alpha} + \Delta q_{C_\alpha}^\circ$.

The hydrogen charge variations are expressed relative to $q_H^\circ = -11.7$ me. Charge normalization, $\sum q_H = -(\sum q_C + \sum q_{C_\alpha} + \sum q_O)$, gives

$$\Delta q_H = -\left(\sum \Delta q_C + \sum \Delta q_{C_\alpha} + \sum \Delta q_O\right)$$
$$-(n_C \, q_C^{\circ C_2 H_6} + n_{C_\alpha} q_{C_\alpha}^\circ + n_O \, q_O^\circ + n_H \, q_H^\circ) \qquad (6.27)$$

where n_{C_α}, n_O, n_C and n_H are the numbers of C atoms adjacent to O, of the O atoms, the C atoms not bonded to O and of the H atoms, respectively.

Ethers

Straightforward summations and (6.27) give the following general formula for dialkylethers:

$$\sum_k \sum_l a_{kl} \Delta q_k = A_1 \sum N_{CC} \delta_C + A_2 \sum \delta_C + A_3 \sum \delta_{C_\alpha} + A_4 \delta_O$$
$$+ \Delta q_C^\circ \left(a_{CC} \sum N_{C_\alpha C} + a_{CH} \sum N_{C_\alpha H} - 2a_{HC}\right)$$
$$+ (n_C - 2)a_{HC} q_H^\circ - a_{HC} q_O^\circ \qquad (6.28)$$

which incorporates the appropriate charge–NMR shift relationships. N_{CC} is the number of CC bonds formed by the carbon whose shift is δ_C. The $\sum N_{CC} \delta_C$ term of (6.28) includes both δ_{C_α} (relative to the diethylether C_α atom) and δ_C (from ethane). $N_{C_\alpha C}$ and $N_{C_\alpha H}$ are, respectively, the number of CC and CH bonds formed by the C_α carbons. The $A_1 = a_{CC} - a_{CH} = 0.0356$ and $A_2 = 4a_{CH} - a_{HC} = 0.0529$ kcal mol^{-1} ppm^{-1} parameters are those found earlier in eqs. (6.11) and (6.22), while A_3 and A_4 are

$$A_3 = 3a_{CH} + a_{CO} - a_{HC}$$
$$A_4 = 2a_{OC} - a_{HC}.$$

Using the appropriate a_{kl} parameters, $a_{CO}(1.43) = -1.135$, $a_{CH}(1.08) = -0.394$ and $a_{HC}(1.08) = -1.007$ au, as well as $\Delta q_C = -0.148\delta_C$ (me), one

TABLE 6.11. Comparison between calculated and experimental atomization
energies of selected ethers, kcal mol^{-1}

Molecule	ΔH_f°	$\sum_{k<l} \varepsilon_{kl}^\circ$	$\sum_{k,l} a_{kl}\Delta q_k$	E_{nb}	ΔE_a^* Calcd.	ΔE_a^* Exptl.
$(CH_3)_2O$	−43.99	800.42	−3.33	−0.05	797.14	797.07
$CH_3OC_2H_5$	−51.72	1083.67	9.56	−0.09	1093.32	1093.07
$CH_3O\text{-}n\text{-}C_3H_7$	−56.82	1366.92	19.35	−0.25	1386.52	1386.43
$CH_3O\text{-}i\text{-}C_3H_7$	−60.24	1366.92	22.82	−0.12	1389.86	1389.85
$(C_2H_5)_2O$	−60.24	1366.92	22.45	−0.17	1389.54	1389.87
$(n\text{-}C_3H_7)_2O$	−69.85	1933.42	41.92	−0.47	1975.81	1976.00
$(i\text{-}C_3H_7)_2O$	−76.20	1933.42	48.97	−0.29	1982.68	1982.35
$(n\text{-}C_4H_9)_2O$	−79.82	2499.92	61.95	−0.73	2562.60	2562.50
$(s\text{-}C_4H_9)_2O$	−86.26	2499.92	68.22	−0.58	2568.72	2568.94
Tetrahydropyran	−53.39	1506.18	50.95	−0.49	1557.62	1557.45
1,4-Dioxane	−75.65	1312.86	39.59	−0.29	1352.74	1352.76

obtains $A_3 = -1.310$ au $= 0.1217$ kcal mol^{-1} ppm^{-1}. Because of the significant variations of Δq_O, one must consider the second order term of (3.35)
in the evaluation of a_{OC}. In practice, it suffices to use $A_4 = 0.1007$ kcal
mol^{-1} ppm^{-1} for $a_{OC} = -0.804$ au ($\delta_O < 0$) or 0.0994 kcal mol^{-1} ppm^{-1}
for $a_{OC} = -0.800$ au ($\delta_O > 0$), where $\Delta q_O = -0.267\delta_O$ me was used [41].

The final energy analyses are displayed in Table 6.11[16]. Equation (6.28)
does not apply to tetrahydropyran and 1,4-dioxane. These molecules were
calculated using the atom-by-atom approach described earlier.

Carbonyl compounds

Summation of the $a_{kl}\Delta q_k$ terms and (6.27) gives the following general formula describing carbonyl compounds RR'CO with R = alkyl and R' = alkyl
or H

$$\sum_k \sum_l a_{kl}\Delta q_k = A_1 \sum N_{CC}\delta_C + A_2 \sum \delta_C + A_3'\delta_{C_\alpha} + A_4'\delta_O$$

$$+ \Delta q_{C_\alpha}^\circ (a_{C_\alpha C}N_{C_\alpha C} + a_{C_\alpha H}N_{C_\alpha H} - a_{HC})$$

$$+ n_C\, a_{HC}q_H^\circ - a_{HC}q_O^\circ \qquad (6.29)$$

[16]The experimental ΔH_f° results are from [24], except that of 1,4-dioxane, taken from
[42]. The ZPE + $H_T - H_0$ results are from Table A.2 and eq. (A.14). The ^{13}C NMR shifts
are from [27] and [43]. The ^{17}O shifts are from [43], except those of tetrahydropyran and
1,4-dioxane which are given in [44].

where

$$A'_3 = 2a_{C_\alpha H} + a_{CO} - a_{HC}$$
$$A'_4 = a_{OC} - a_{HC}$$

The other parameters of (6.29) are those explained for the dialkylethers. $\sum N_{CC}\delta_C$ includes both δ_{C_α} (from the acetone carbonyl-C atom) and δ_C (from ethane) for the C atoms not bonded to O. Selected a_{kl} values are [3, 41] $a_{C_\alpha H}(1.08) = -0.276$ and $a_{CO}(1.22) = -1.182$ au. Hence $A'_3 = -0.727$ au. Tentatively assuming $\Delta q_C = -0.148\delta_C$ me [3, 41], we get $A'_3 = 0.0675$ kcal mol^{-1} ppm^{-1}. Noting that $N_{C_\alpha C} + N_{C_\alpha H} = 2$, we rewrite (6.29) as follows

$$\sum_k \sum_l a_{kl}\Delta q_k = 0.0356 \sum N_{CC}\delta_C + 0.0529 \sum \delta_C + 0.0675\delta_{C_\alpha} + A'_4\delta_O$$
$$+7.393n_C + 5.07N_{C_\alpha C} - 19.42 \text{ kcal mol}^{-1} \qquad (6.30)$$

where n_C is the total number of C atoms. Here we cannot take a_{OC} as a constant because of the significant variations of Δq_O. Using the γ's of Table 6.1 and the appropriate first and second derivatives of Table 6.2, one obtains from (3.35) that $a_{OC}(1.22) = -1.065-0.254\Delta q_O$ and $A'_4 = -(0.058+ 0.254\Delta q_O)$ au. Using $\Delta q_O(\text{carbonyl}) \simeq 2.7\delta_O$ me [3, 41], we get

$$A'_4 \simeq -(0.098 + 1.16 \times 10^{-3} \times \delta_O) \text{ kcal mol}^{-1} \text{ ppm}^{-1}$$

The validity of this approximation is best illustrated by the results offered in Table 6.12[17]. Surely, eq. (6.30) permits accurate calculations of atomization energies and represents a simple and valuable tool. It must be made clear, however, that the charge–NMR shift correlations used for the atoms of the carbonyl group—i.e., those involved in the $A'_3\delta_{C_\alpha}$ and $A'_4\delta_O$ terms— are entirely empirical. On the other hand, it turns out [3, 41] that for the ketones $A'_3\delta_{C_\alpha} + A'_4\delta_O$ amounts to less than 5% of $\sum_k \sum_l a_{kl}\Delta q_k$. Hence, with some reservations in mind, it seems reasonable to claim that at least the main features of the theory underlying these calculations withstand challenging tests like those offered in Table 6.12.

---

[17]The experimental ΔH_f° results are extracted from [24]. As concerns the ZPE + $H_T - H_0$ energies [3], an increment of 18.3 kcal mol^{-1} was assumed for each added CH$_2$ group with respect to the closest 'parent' compound, in line with the results obtained for the parent hydrocarbons [45]; see also Table A.2. The ^{13}C NMR shifts are reported in [27, 46] and the ^{17}O shift are taken from [46]. The nonbonded contributions, calculated from (3.11), are given in [41]; for carbonyl compounds, they should be interpreted with some reservations and are used on a tentative basis.

TABLE 6.12. Comparison between calculated and experimental atomization energies of selected carbonyl compounds, kcal mol^{-1}

Molecule	ΔH_f°	$\sum \varepsilon_{kl}^\circ$	$\sum_{k,l} a_{kl}\Delta q_k$	E_{nb}	ΔE_a^* Calcd.	ΔE_a^* Exptl.
CH_3CHO	-39.73	676.25	-0.78	-0.30	675.77	675.69
C_2H_5CHO	-45.45	959.50	10.17	-0.36	970.03	970.34
$n\text{-}C_3H_7CHO$	-48.98	1242.75	19.26	-0.75	1262.76	1262.76
$(CH_3)_2CO$	-51.90	959.50	16.85	-0.07	976.42	976.43
$CH_3COC_2H_5$	-57.02	1242.75	27.19	-0.45	1270.39	1270.48
$CH_3CO\text{-}n\text{-}C_3H_7$	-61.92	1526.00	37.52	-0.42	1563.94	1564.30
$CH_3CO\text{-}i\text{-}C_3H_7$	-62.76	1526.00	38.75	-0.55	1565.30	1565.14
$(C_2H_5)_2CO$	-61.65	1526.00	37.42	-0.92	1564.34	1564.36
$CH_3CO\text{-}n\text{-}C_4H_9$	-66.96	1809.25	48.82	-0.55	1858.62	1858.27
$CH_3CO\text{-}tert\text{-}C_4H_9$	-69.28	1809.25	51.44	-0.6	1861.3	1860.8
$C_2H_5CO\text{-}n\text{-}C_3H_7$	-66.51	1809.25	48.11	-1.05	1858.41	1858.15
$C_2H_5CO\text{-}i\text{-}C_3H_7$	-68.38	1809.25	49.40	-1.05	1859.70	1859.70
$C_2H_5CO\text{-}tert\text{-}C_4H_9$	-74.99	2092.50	61.49	-1.2	2155.2	2155.2
$(i\text{-}C_3H_7)_2CO$	-74.40	2092.50	60.60	-1.85	2154.95	2154.97

6.8 Concluding comments

The numerical tests presented in Sections 5–7 can be rated with respect to the quality of the input data, primarily the charges, used in the calculations. The alkanes certainly rank highest in the hierarchy of confidence because their charges leave no room for empirical 'readjustments' except, perhaps, as regards minor uncertainties surrounding the absolute value assumed for the reference net charge of the ethane carbon atom. At any rate, it is clear that the charges described in Chapter 5 are the only ones fit for use in our energy formulas.

The same confidence is certainly justified in energy calculations of the alkyl parts of organic molecules containing functional groups, such as a carbon–carbon double bond, a carbonyl group, etc. In contrast, our knowledge about the charges of functional groups, e.g., those involving sp^2 carbon atoms, is not as solid as desired. Luckily, it turns out that the terms depending on the lesser known charges are numerically considerably less important than the largely preponderant part of the calculation that is firmly known, thus reducing the impact of the blur introduced into the final results. It seems fair to claim that the tests presented for the unsaturated hydrocarbons adequately support the bond-energy theory. With the oxygen-containing molecules, however, all we can presently demonstrate is that the assumptions made for the charges of the functional groups, C–O and C=O, lead to excellent final energies but, because of these assumptions, these tests rank rather low as 'verification' of our energy formulas.

We have learned that an adequate knowledge about atomic charges is a must in applications of our bond-energy formulas and that accurate charge–^{13}C NMR shift correlations offer valuable guidelines as to the plausibility of calculated charges. We have also learned that saturated hydrocarbons represent a class of molecules best suited for illustrating the combination of these two facets of theory: first, a most convenient access to reliable charges and, second, their use in energy calculations. There is an important restriction, however, that must be strongly pointed out. Charge analyses and, in particular, charge–^{13}C NMR shift correlations valid for normal and branched paraffins and six-membered chair or boat rings are not applicable to small-ring cycloalkanes. Too little is presently known about these systems to permit a lucid discussion, but work on cyclopropane [47] and tetracyclo[4.1.0.0^{2,4}.0^{3,5}]heptane [48] unmistakably demonstrates that three-membered rings cannot be treated like six-membered cycloalkanes—but this should not come as a surprise. Future work should certainly consider SDCI correlation corrections in charge analyses of strained rings. Presently, the only indication we have is that the energy of each CC bond in cyclopropane roughly equals that of ethane itself [16]. Similar charge–shift ring-size effects are also observed with azulene [49] and await additional investigations.

Correlations between nuclear magnetic resonance shifts and atomic charges of nitrogen in selected alkylamines, nitroalkanes, isonitriles and azines consistently follow the general trends observed for carbon and oxygen nuclei [50]. Future energy calculations of nitrogen-containing molecules could benefit from these correlations if the following questions can be answered. The first question concerns alkylamines. The slope of the ^{15}N chemical shifts *vs.* the nitrogen net charges observed for tertiary amines differs significantly from that found for the mono- and dialkylamines. Is this a computational artefact linked to basis set superposition errors or is this change of slope simply due to solvent and concentration effects in the measurements of chemical shifts? The next question concerns hydrogen net charges. In the calculation of molecules where H atoms can be attached either to carbon or to nitrogen, as in the mono- and dialkylamines, for example, it becomes necessary to know them individually or, at least, the total charge of the hydrogen atoms attached to nitrogen (or to carbon). Surely, these problems can be solved, thus permitting future energy calculations of nitrogen-containing molecules.

Bibliography

[1] C. Froese Fischer, *At. Data Nucl. Data Tables*, **4**, 301 (1972); *ibid.*, **12**, 87 (1973).

[2] P. Politzer, *J. Chem. Phys.*, **70**, 1067 (1979).

[3] S. Fliszár, "Charge Distributions and Chemical Effects", Springer-Verlag, New York, 1983.

[4] T. Anno, *J. Chem. Phys.*, **72**, 782 (1980).

[5] S. Fliszár and M.-T. Béraldin, *J. Chem. Phys.*, **72**, 1013 (1980).

[6] S. Fliszár, M. Foucrault, M.-T. Béraldin, and J. Bridet, *Can. J. Chem.*, **59**, 1074 (1981).

[7] P. Politzer, *J. Chem. Phys.*, **64**, 4239 (1976).

[8] K. Ruedenberg, *J. Chem. Phys.*, **66**, 375 (1977).

[9] J. C. Slater, *Adv. Quantum Chem.*, **6**, 1 (1972); J. C. Slater, *Int. J. Quantum Chem. Symp.*, **3**, 727 (1970); J. C. Slater *in* "Computational Methods in Band Theory", P. M. Marcus, J. F. Janak and A. R. Williams (eds.), Plenum Press, New York, London, 1971, p 447; J. C. Slater, "The Self-Consistent Field for Molecules and Solids", vol. 4, McGraw-Hill, Kuala-Lumpur, 1974.

[10] T. H. Dunning, *J. Chem. Phys.*, **53**, 2823 (1970).

[11] K. Schwarz, *Phys. Rev.*, **B5**, 2466 (1972).

[12] S. Fliszár, G. Cardinal, and N. A. Baykara, *Can. J. Chem.*, **64**, 404 (1986).

[13] S. Fliszár, *J. Am. Chem. Soc.*, **102**. 6946 (1980).

[14] W. J. Hehre, L. Radom, P. v.R. Schleyer, and J. A. Pople, "Ab initio Molecular Orbital Theory", John Wiley & Sons, New York, 1986.

[15] L. S. Bartell, K. Kuschitsu, and R. J. DeNui, *J. Chem. Phys.*, **35**, 1211 (1961); D. E. Shaw, D. W. Lepard, and H. L. Welsh, *J. Chem. Phys.*, **42**, 3736 (1965); K. Kuschitsu, *J. Chem. Phys.*, **44**, 906 (1966).

[16] V. Barone and S. Fliszár, to be published.

[17] B. P. Stoicheff, *Can. J. Phys.*, **32**, 339 (1954).

[18] J. Kao and N. L. Allinger, *J. Am. Chem. Soc.*, **99**, 975 (1977); A. Almenningen and O. Bastiansen, *Skr. K. Nor. Vidensk. Selsk.*, **4**, 1 (1958).

[19] S. Fliszár and M.-T. Béraldin, *Can. J. Chem.*, **57**, 1772 (1979).

[20] H. Henry and S. Fliszár, *J. Am. Chem. Soc.*, **99**, 5889 (1977).

[21] J. L. Marshall and E. D. Canada, Jr., *J. Org. Chem.*, **45**, 3124 (1980).

[22] T. Clark, T. McO. Knox, H. Mackle, A. McKervey, and J. J. Rooney, *J. Am. Chem. Soc.*, **97**, 3835 (1975).

[23] "Sadtler Standard Carbon-13 NMR Spectra", Sadtler Research Laboratories Inc., Philadelphia, PA.

[24] J. D. Cox and G. Pilcher, "Thermochemistry of Organic and Organometallic Compounds", Academic Press, New York, NY, 1970.

[25] M.-T. Béraldin and S. Fliszár, *Can. J. Chem.*, **61**, 197 (1983).

[26] S. W. Benson, F. R. Cruickshank, D. M. Golden, G. R. Haugen, H. E. O'Neal, A. S. Rodgers, R. Shaw, and R. Walsh, *Chem. Rev.*, **69**, 279 (1969).

[27] J. B. Stothers, "Carbon-13 NMR Spectroscopy", Academic Press, New York, NY, 1972.

[28] G. J. Abruscato, P. D. Ellis, and T. T. Tidwell, *J. Chem. Soc., Chem. Comm.*, 988 (1972).

[29] J. W. de Haan, L. J. M. van der Ven, A. R. N. Wilson, A. E. van der Hout-Lodder, C. Altona, and D. H. Faber, *Org. Magn. Res.*, **8**, 477 (1976).

[30] J. W. de Haan and L. J. M. van der Ven, *Org. Magn. Res.*, **5**, 147 (1973).

[31] S. Fliszár and G. Cardinal, *Can. J. Chem.*, **62**, 2748 (1984).

[32] T. H. Dunning, Jr., *J. Chem. Phys.*, **90**, 1007 (1989); R. A. Kendall and T. H. Dunning, Jr., *J. Chem. Phys.*, **96**, 6796 (1992); D. E. Woon and T. H. Dunning, Jr., *J. Chem. Phys.*, **98**, 1358 (1993).

[33] R. Munday and I. O. Sutherland, *J. Chem. Soc.* B, 80 (1968).

[34] P. R. Pinnock, C. A. Taylor, and H. Lipson, *Acta Crystallogr.*, **9**, 173 (1956); F. R. Ahmed and J. Trotter, *Acta Crystallogr.*, **16**, 503 (1963).

[35] M. J. S. Dewar and C. de Llano, *J. Am. Chem. Soc*, **91**, 789 (1969).

[36] D. E. Gray (ed.), "American Institute of Physics Handbook", 3rd ed., McGraw-Hill, New York, NY, 1972, p. 905.

[37] A. Peluso and S. Fliszár, *Can. J. Chem.*, **66**, 2631 (1988).

[38] ICSU-CODATA Task Group on Key Values for Thermodynamics, *J. Chem. Thermodyn.*, **3**, 1 (1971).

[39] D. D. Richardson, *J. Phys. C. Sol. State Phys.*, **10**, 3235 (1977); E. Santos and A. Villagrá, *Phys. Rev.*, **B6**, 3134 (1972); R. J. Good, L. A. Girifalco, and G. Kraus, *J. Phys. Chem.*, **62**, 1418 (1958).

[40] JANAF Thermochemical Tables, 3rd ed., *J. Phys. Chem. Ref. Data*, **14**, Suppl. 1, 536 (1985).

[41] S. Fliszár and M.-T. Béraldin, *Can. J. Chem.*, **60**, 792 (1982).

[42] K. Byström and M. Månsson, *Perkin Trans.*, **2**, 565 (1982).

[43] C. Delseth and J.-P. Kintzinger, *Helv. Chim. Acta*, **61**, 1327 (1978).

[44] T. Sugawara, Y. Kawada, M. Katoh, and H. Iwamura, *Bull. Chem. Soc. Japan*, **52**, 3391 (1979).

[45] S. Fliszár and J.-L. Cantara, *Can. J. Chem.*, **59**, 1381 (1981).

[46] C. Delseth and J.-P. Kintzinger, *Helv. Chim. Acta*, **59**, 466 (1976); *ibid.*, **59**, 1411 (1976).

[47] R. Roberge and S. Fliszár, *Can. J. chem.*, **53**, 2400 (1975).

[48] J. Bridet, M.-T. Béraldin, and S. Fliszár, *Can. J. Chem.*, **63**, 2468 (1985).

[49] S. Fliszár, G. Cardinal, and M.-T. Béraldin, *J. Am. Chem. Soc.*, **104**, 5287 (1982).

[50] M. Comeau, M.-T. Béraldin, E. Vauthier, and S. Fliszár, *Can. J. Chem.*, **63**, 3226 (1985).

Chapter 7

Assessment

In these *Lecture Notes*, chemical bonds, their intrinsic energies and the energies required for bond breaking are at the very center of our attention. Rooted in Hartree–Fock theory, our approach ends up with a formalism, reminiscent of that given by Thomas–Fermi theory, featuring the electrostatic potentials at the nuclei. Both for ground-state atoms and molecules at equilibrium in the Born–Oppenheimer approximation, the total (kinetic + potential) energies are expressed as simple functions of these potentials. The latter are obtained from straightforward applications of the Hellmann–Feynman theorem, the nuclear charges being taken as parameters in these energy derivatives. For molecules, considering such an energy derivative for each individual nucleus, we thus introduce the concept of 'atom in a molecule'—differing, of course, from an isolated ground-state atom—and then, from an appropriate pairing of atomic terms, we achieve an effective bond-by-bond partitioning of a molecule—*avec tout ce que cela comporte*—carrying a vivid operational mathematical representation of chemical bonds and their intrinsic energies in molecules at equilibrium.

The simplifying reduction of the general problem of molecular energies—and thus of bond energies—to its electrostatic aspects entails the loss of the reasons explaining *why* chemical bonds are formed in the first place: it is well known that the mechanism of bond formation cannot be understood in purely electrostatic terms because the electronic Hamiltonian of an atom or a molecule is only bounded from below if the kinetic energy is correctly accounted for. Beyond this restriction, however, our approach is surely well suited for describing important properties of chemical bonds in molecules at equilibrium, namely, their energies.

The theory begins with a description of isolated atoms; more precisely: with a description of their core and valence regions. Why? The answer is in the formula for the energy, E^v, of the outer valence electrons, namely

$$E^v = -constant \times (Z - N^c) \int_{r_b}^{\infty} \frac{\rho(\mathbf{r})}{r} d\mathbf{r}$$

where the '*constant*' is $\sim \frac{3}{7}$, as in the Thomas–Fermi model. This equation features the potential due to the cloud of valence electrons: the point is that *this 'outer' electronic charge must be considered in the field of an 'effective' nucleus*, $Z^{eff} = Z - N^c$, partially screened by its core electrons. A similar situation arises with the electrostatic potential at the nucleus Z_k of an atom k found in a molecule. The potential contributed by the electronic and nuclear charge of the other atoms of that molecule is assimilated to a potential due to 'outer' charges (with respect to $Z_k - N_k^c$) and must be taken in the field of the effective nuclear charge $Z_k^{eff} = Z_k - N_k^c$ of atom k. Briefly, it turns out that our Thomas–Fermi-like treatment of molecules presupposes a description of valence-region electrons in the field of effective nuclear charges Z_k^{eff}, Z_l^{eff}, etc. In this presentation of the theory, it is now shown how the core–valence separation in molecules is a simple follow-up of that developed for atoms— this is why we begin things with isolated atoms—and how it offers a route towards the theory of chemical binding. Note that the transformation of the pertinent Hartree–Fock formulas to give the appropriate, more manageable Thomas–Fermi-like energy expressions implies acceptance of Gauss' theorem for the 'inner' core electrons, which seems to be a reasonable assumption.

The final energy formula for a chemical bond

$$\varepsilon_{kl} = \varepsilon_{kl}^{\circ} + a_{kl}\Delta q_k + a_{lk}\Delta q_l$$

translates intuitive expectations, namely, that the intrinsic energy of a bond formed between atoms k and l should depend on the charges carried by these atoms. The parameters (i.e., $\varepsilon_{kl}^{\circ}, a_{kl}$ and a_{lk}) are given by theory. The direct theoretical evaluation of ε_{kl}°, the reference bond energy corresponding to reference charges q_k° and q_l°, respectively, turns out to be particularly difficult, but, in actual practice, only few reference bonds need be calculated to give access to large series of homologous molecules. There is a hidden variable in this formulation of ε_{kl}, namely, pertinent information regarding effects linked to the shape of valence electron populations or, more precisely, to the centroids of valence atomic orbitals. Simple approximations allow us to deal efficiently with problems of this sort and to incorporate the required corrections into the ε_{kl}° term. Now, our sole concern is about charges.

151

We know, of course, that the characterization of physically meaningful atomic charges and the description of bond energies proceed from theories that are entirely independent of one another: the bond theory does not 'shape' in any way that of the charges occurring in it. On the other hand, we also know that realistic atomic charges can be defined for LCAO-MO wave functions, applying a modified Mulliken population analysis, although the general solution of this partitioning problem has yet to be found. For the saturated hydrocarbons, however, we are able to come up with a unique solution. It is rewarding—and certainly most encouraging—to witness that these are precisely the charges that must be used in applications of our bond-energy formulas. They ought to be remembered: they provide valuable guidelines and criteria permitting to assess the merits of present and future charge partitioning schemes.

Individual bond energies cannot be checked against experimental values, but their sums, $\sum_{k<l} \varepsilon_{kl}$, nearly equal the appropriate molecular atomization energies and are thus amenable to experimental verifications. The summary of tests presented in Figure 7.1 is surely convincing because calculations of atomization energies are extremely sensitive to any malfunctioning of charge analyses. The charges used here, though not perfect, are certainly adequate, particularly as far as their breakdown in alkyl groups is concerned.

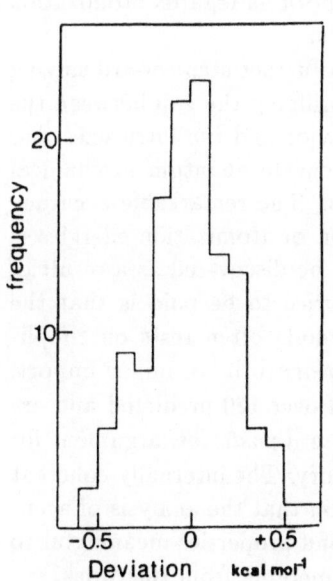

FIGURE 7.1. Frequency of deviations between calculated and experimental atomization energies. This error distribution survey covers alkanes (including cycloalkanes constructed from chair and/or boat cyclohexane rings), simple ethylenic and polyunsaturated nonaromatic hydrocarbons, as well as ethers, aldehydes and ketones. The average deviation is $\sim 0.23\,\text{kcal mol}^{-1}$! Similar results were also obtained for ~ 40 benzenoid hydrocarbons, including graphite. (Reproduced from S. Fliszár, G. Cardinal, and N.A. Baykara, *Can. J. Chem.*, **64**, 404, 1986.)

Everything adds up nicely at this point, at least in principle. Our bond-energy theory—and its numerical applications—sheds light on the relevant electronic effects governing the fine-tuning of molecular energies and thus visualizes the subtle physics explaining energy differences between structural isomers, dominated by bond-energy changes driven by changing electron distributions. Briefly, things are well under control as long as we calculate sums of bond energies to be compared with atomization energies.

What about the individual bonds? Here we must make an important distinction between intrinsic bond energies—which give information about the thermochemical stability of a molecule—and bond dissociation energies, which mirror properties related to the reactivity of that molecule.

Intrinsic bond energies can be calculated but cannot be measured. Bond dissociation energies, on the other hand, can be measured but cannot be predicted from the ground-state properties of the molecule where the bonds occur. Well, that's odd, but this deadlock can be avoided with the help of a relationship existing between intrinsic bond energies, ε_{kl}, and bond dissociation energies, D_{kl}, a relationship easily deduced from first-principle energy and electronic charge conservation constraints. Just like ε_{kl}, D_{kl} depends on the charges carried by atoms k and l, a charge dependence clearly supported by pertinent experimental results. Briefly, it seems fair to claim that the basic tenets underlying our bond-energy theory are reasonably well established and supported by experimental results, both as regards atomization energies and the dissociation of individual bonds.

This bond-energy theory is to some extent a further step toward solving a persistent problem of theoretical chemistry: bridging the gap between the apparent simplicity of observed molecular behavior and the intricacies and ambiguities which plague the translation of accurate quantum mechanical results into simple chemical concepts and rules. The remarkable accuracy with which our energy formulas allow prediction of atomization energies is a strong indication that hidden regularities can be discovered as a result of patient processing of general theory, but the price to be paid is that the use of final equations like our bond-energy formula often rests on simplifying assumptions which cannot be proven *a priori* to be of minor import. The agreement within experimental accuracy of over 170 predicted and experimental atomization energy values provides an *a posteriori* argument for claiming that these approximations are satisfactory. The internally coherent picture thus obtained provides a strong indication that the analysis of accurate *ab initio* computations into models, rules, and properties meaningful to chemists should be carried out along the lines emerging from this work.

Appendix A

The Comparison with Experiment

This Appendix offers information facilitating comparisons between theory and experiment using thermochemical data. Special attention is given to the evaluation of zero-point and heat-content energies which are often not as readily available as desired.

A.1 Thermochemical formulas

The atomization

$$\text{Molecule} \longrightarrow n_1 A_1 + n_2 A_2 + \cdots + n_k A_k$$

of a molecule into its constituent n_1 atoms A_1, n_2 atoms A_2, etc. provides a measure for chemical binding. The relevant thermodynamic information is usually expressed in terms of enthalpy of formation, ΔH_f, or enthalpy of atomization, ΔH_a, of the molecule under consideration, i.e.,

$$\Delta H_a = \sum_k n_k \Delta H_f(A_k) - \Delta H_f \tag{A.1}$$

where the $\Delta H_f(A_k)$'s are the enthalpies of formation of the gaseous atoms A_k. In the study of isolated molecules, however, we refer more appropriately to the *energy* of atomization, ΔE_a, that is

$$\Delta E_a = \Delta H_a - \left(\sum_k n_k - 1 \right) RT. \tag{A.2}$$

Both ΔE_a and ΔH_a are considered at some temperature, usually 25°C, i.e., under working conditions of practical interest. However, when used in the study of molecular properties, these quantities contain the seed of unnecessary difficulties, such as those arising from internal rotations which are more or less free in some cases and hindered in others. This is avoided by studying the molecules at 0 K. However, zero-point vibrational energies are to be taken into account, as these energies, like the thermal ones, cannot fairly be apportioned among bonds or atoms in a molecule since they are not truly additive properties nor can they be regarded as a part of chemical binding. These reasons prompt us to study a molecule in its hypothetical vibrationless state at 0 K, whose atomization energy is ΔE_a^*. This is the ΔE_a^* of eq. (3.2).

The relationship between ΔE_a^* and ΔE_a is described by eq. (A.3)

$$\Delta E_a = \Delta E_a^* - \sum_i f(\nu_i, T) + \frac{3}{2}\left(\sum_k n_k - 2\right) RT \qquad (A.3)$$

which states that the energy of atomization at, say, 25°C is that of the hypothetical vibrationless molecule at 0 K *less* the sum $\sum_i f(\nu_i, T)$ (over $3\sum_k n_k - 6$ degrees of freedom) of vibrational energy corresponding to the fundamental frequencies ν_i, which is already present in the molecule at 25°C. The term $3(\sum_k n_k - 2)RT/2$ accounts for the formation of $\sum_k n_k$ atoms with translational energy and the disappearance of one nonlinear molecule with 3 translational and 3 rotational $RT/2$ contributions. It follows from eqs. (A.1)–(A.3) that

$$\Delta E_a^* = \sum_k n_k \left[\Delta H_f^\circ(A_k) - \frac{5}{2}RT\right] + \sum_i f(\nu_i, T) + 4RT - \Delta H_f^\circ \qquad (A.4)$$

where all enthalpies are now referred to standard conditions (gas, 298.15 K). The vibrational energy may be separated into a zero-point energy term (ZPE) and a thermal vibrational energy term, E_{therm}. It is therefore

$$\sum_i f(\nu_i, T) + 4RT = ZPE + E_{therm} + 4RT. \qquad (A.5)$$

On the other hand, $E_{therm} + 4RT$ is the increase in enthalpy $(H_T - H_0)$ of nonlinear molecules due to their warming up from $T = 0$ to $T = T$. Consequently, eq. (A.4) can now be written as follows

$$\Delta E_a^* = \sum_k n_k \left[\Delta H_f^\circ(A_k) - \frac{5}{2}RT\right] + ZPE + (H_T - H_0) - \Delta H_f^\circ. \qquad (A.6)$$

The same formula applies to linear molecules, but then $(H_T - H_0) = E_{therm} + \frac{7}{2}RT$. This is the basic equation permitting a comparison between calculated ΔE_a^* results and thermochemical information. The appropriate standard enthalpies of formation of the atoms, $\Delta H_f^\circ(A_k)$, are [1] $\Delta H_f^\circ(C) = 170.89$, $\Delta H_f^\circ(H) = 52.09$, and $\Delta H_f^\circ(O) = 59.54$ kcal mol^{-1} (gas, 298.15 K).

Zero-point energies are obtained from vibrational spectra using experimental frequencies whenever available, while the inactive frequencies are extracted from data calculated by means of an appropriate force-field model. In the harmonic oscillator approximation, the zero-point energy is

$$ZPE = \frac{1}{2} \sum_i h\nu_i \qquad (A.7)$$

and the thermal vibrational energy is obtained from Einstein's equation

$$E_{therm} = \sum_i h\nu_i \left(\frac{\exp(-h\nu_i/kT)}{1 - \exp(-h\nu_i/kT)} \right) \qquad (A.8)$$

from which $(H_T - H_0) = E_{therm} + 4RT$ is readily deduced. The infrared and Raman fundamental frequencies are usually expressed as wave numbers, in cm^{-1} units. Taking Planck's constant at $h = 6.6256 \times 10^{-27}$ erg sec, Boltzmann's constant at $k = 1.38054 \times 10^{-16}$ erg K^{-1} and $c = 2.997925 \times 10^{10}$ cm s^{-1} for the speed of light, it is for wave numbers ω_i expressed in cm^{-1}:

$$\frac{1}{2}h\nu_i = 1.42956 \times 10^{-3}\omega_i, \quad \text{kcal mol}^{-1}$$

$$h\nu_i/kT = 4.82572 \times 10^{-3}\omega_i, \quad \text{at 298.15K.}$$

A.2 Zero-point and heat-content energies

Equation (A.6) indicates how to calculate the energy of atomization of a molecule at its potential minimum from thermochemical and spectroscopic information. The required standard enthalpies of formation are known experimentally for large collections of molecules. The zero-point plus heat-content energies, $ZPE + (H_T - H_0)$, can be calculated from the complete sets of fundamental vibrational frequencies. Everything seems fine, except for the unfortunate fact that resolved complete vibrational spectra are usually not as readily available as desired, thus rendering *bona fide* comparisons between theory and experiment difficult, in spite of a wealth of thermochemical information. Hence our interest in alternate rules permitting the construction of reliable vibrational energies in a simple manner, thus facilitating the use of eq. (A.6).

It is fortunate that ZPE + $(H_T - H_0)$ energies obey, to a good approximation, a number of simple additivity rules [2]. Theoretical foundation for this additivity, which has a long history [3], has been established [4]. In the following, we examine ZPE + $(H_T - H_0)$ energies deduced from spectroscopic data and relate these results to structural features. In essence, this is a presentation of facts—a point worth remembering because, at times, the observed regularities exceed present theoretical expectations.

The results are derived in the harmonic oscillator approximation using whenever possible observed infrared and Raman fundamental frequencies. The inactive frequencies are taken from appropriate force-field calculations. All the results are for $T = 298.15$ K. The results cover i) open-chain alkanes, alkenes and alkynes, ii) open-chain molecules containing heteroatoms, iii) non-aromatic cyclic molecules, iv) aromatic molecules and v) free radicals.

Open-chain alkanes, alkenes and alkynes

A detailed study of the alkanes has revealed [5] that their ZPE + $(H_T - H_0)$ energies are satisfactorily described as follows, in kcal mol^{-1},

$$ZPE + (H_T - H_0) = 11.479 + 18.213n_C - 0.343n_{br} \qquad (A.9)$$

where n_C is the number of carbon atoms and n_{br} is the number of branchings. The average deviation between the results deduced in this manner and their spectroscopic counterparts is 0.125 kcal mol^{-1} (~ 80 cm^{-1}). The change in ZPE + $(H_T - H_0)$ associated with methyl substitutions giving quaternary carbon atoms, however, can only be estimated as a rough average, $\sim 17.55 \pm 0.25$ kcal mol^{-1}, and should be considered with caution.

For the simple ethylenic compounds, C_nH_{2n}, it was shown [2] that eq. (A.10), in kcal mol^{-1},

$$ZPE + (H_T - H_0) = 33.35 + 18.213(n_C - 2) - 0.343n_{br} \qquad (A.10)$$

represents a valid approximation. (Situations of extreme steric crowding, such as those arising with the presence of two *tert*-butyl groups attached to the same carbon, are not represented by this equation.) Now we proceed with conjugated and nonconjugated polyenic hydrocarbons. Their ZPE + $(H_T - H_0)$ energies can be estimated [6] from those of their olefinic fragments, eq. (A.10), simply bt subtracting 11.56 kcal mol^{-1} for each pair of hydrogen atoms eliminated in the condensation of the fragments. For example, the result predicted for 1,3- pentadiene follows from the values of ethene and propene less 11.56 kcal mol^{-1}. Similarly, the result for 1,3,5-hexatriene corresponds to three times that of ethene less twice 11.56 kcal mol^{-1} or, alternatively, to the sum obtained from ethene and butadiene less 11.56 kcal mol^{-1}. Alternatively, taking now the experimental value of butadiene (55.19 kcal mol^{-1}) as reference, we can describe the dienes as follows [7], in kcal mol^{-1},

$$ZPE + (H_T - H_0) = 55.19 + 18.213(n_C - 4) - 0.343 n_{br} \qquad (A.11)$$

While this type of estimate usually carries an uncertainty not exceeding ~ 0.2 kcal mol^{-1}, it remains that for the dienes (as for the mono-olefins) no spectroscopic information is presently available that would discriminate ZPE $+(H_T - H_0)$ energies of *cis* and *trans* isomers in a reliable manner.

The alkynes are described [6] by an equation similar to that given for the alkenes, namely, in kcal mol^{-1},

$$ZPE + (H_T - H_0) = 18.74 + 18.13(n_C - 2) - 0.343 n_{br} \qquad (A.12)$$

with an average error of ~ 0.14 kcal mol^{-1}.

Inspection of eqs. (A.9)–(A.12) indicates that the addition of one CH_2 group increases ZPE $+ (H_T - H_0)$ by ~ 18.2 kcal mol^{-1}. In addition, the values for ethane (47.91), ethene (33.35) and ethyne (18.74 kcal mol^{-1}), froms eqs. (A.9), (A.10) and (A.12), respectively, indicate that ZPE$+(H_T - H_0)$ decreases regularly by ~ 14.6 kcal mol^{-1} for each added multiple bond. Thus we can describe the alkanes, alkenes and alkynes by a single equation, namely, in kcal mol^{-1},

$$ZPE + (H_T - H_0) = 47.77 + 18.30(n_C - 2) - 0.474 n_{br} - 14.605 n_{mb} \quad (A.13)$$

where n_{mb} is the number of multiple bonds (e.g., 1 in C_2H_4, 2 in C_2H_2). The validity of this approximation is illustrated in Table A.1[1]. The average deviation, 0.20 kcal mol^{-1}, is only slightly larger than what is obtained from separate correlations, confined to series of homologous molecules (identified under the heading A in Table A.1). The striking feature lies in the remarkably regular decrease of vibrational energy in going from a single to a double and to a triple carbon-carbon bond.

---

[1]The references indicate the source of the fundamental frequencies. The $(H_T - H_0)$ results for olefins are from [8].

TABLE A.1. Comparison between experimental and predicted
ZPE + $(H_T - H_0)$ energies (kcal mol^{-1}) of hydrocarbons

| Molecule | ZPE + $(H_T - H_0)$ | | |
	A	(A.13)	Exptl.
Methane	29.69	29.47	29.50	[9]
Ethane	47.91	47.77	48.02	[9]
Propane	66.12	66.07	65.94	[9]
Butane	84.33	84.37	84.38	[9]
Isobutane	83.99	83.90	83.89	[9]
Pentane	102.54	102.67	102.83	[9]
Isopentane	102.20	102.20	102.14	[9]
Hexane	120.76	120.97	120.99	[9]
2,3-Dimethylbutane	120.07	120.02	119.65	[9]
Heptane	138.97	139.27	139.29	[9]
Ethene	33.35	33.17	33.36	[10]
Propene	51.56	51.47	51.82	[11]
1-Butene	69.78	69.77	69.73	[12]
cis-2-Butene	69.78	69.77	69.91	[11]
trans-2-Butene	69.78	69.77	69.70	[11]
Isobutene	69.44	69.29	69.57	[11]
1,3-Butadiene	55.14	55.16	55.19	[13]
trans-1,3-Pentadiene	73.35	73.46	73.28	[14]
cis-1,3-Pentadiene	73.35	73.46	73.32	[14]
Isoprene	73.01	72.99	72.85	[15]
Dimethyl-1,3-butadiene	90.88	90.81	91.08	[16]
trans-1,3,5-Hexatriene	76.93	77.16	76.70	[17]
cis-1,3,5-Hexatriene	76.93	77.16	76.99	[17]
trans,trans-1,3,5,7-Octatetraene	98.72	99.15	99.97	[18]
Ethyne	18.74	18.56	18.57	[19]
Propyne	36.87	36.86	37.05	[19]
1-Butyne	55.00	55.16	54.98	[20]
2-Butyne	55.00	55.16	55.17	[19]
1-Pentyne	73.13	73.46	72.98	[20]
Allene		36.86	36.20	[19]

TABLE A.2. ZPE + $(H_T - H_0)$ energies of carbonyl compounds and ethers

Molecule	ZPE + $(H_T - H_0)$		
	Exptl.	Estd.	
CH_3CHO	36.65		[21]
$(CH_3)_2CO$	54.59		[21]
$(C_2H_5)_2CO$	91.52		[22]
$(CH_3)_2O$	52.65	52.55	[23]
$CH_3OC_2H_5$	69.86	70.19	[24]
$CH_3Oi\text{-}C_3H_7$	87.66	87.83	[24]
$(C_2H_5)_2O$	87.76	87.83	[24]
$C_2H_5Oi\text{-}C_3H_7$	105.40	105.47	[24]
$(i\text{-}C_3H_7)_2O$	123.32	123.11	[24]

Open-chain molecules containing heteroatoms

Complete vibrational analyses of carbonyl compounds are particularly scarse. Results obtained from experimental and calculated fundamental frequencies of acetaldehyde, acetone and diethylketone are indicated in Table A.2. An extensive analysis [2] involving theoretical ΔE_a^* energies and experimental enthalpies of formation, eq. (A.6), indicates that an increment of ~18.3 kcal mol^{-1} can be associated with each added CH_2 group, with respect to the closest parent compound whose spectroscopic ZPE + $(H_T - H_0)$ result is known—which seems a reasonable approximation in light of eq. (A.13). For example, the value for propanal (36.65 + 18.3) is estimated from that of ethanal (36.65 kcal mol^{-1}) and the value for butanone (72.89) is estimated from that of propanone (54.59 kcal mol^{-1}).

The ZPE + $(H_T - H_0)$ results for dialkylethers [2] are adequately represented by the approximation shown in (A.14), in kcal mol^{-1},

$$ZPE + (H_T - H_0) = 52.55 + 17.64(n_C - 2) \qquad (A.14)$$

with an average deviation between predicted and experimental results of ~ 0.16 kcal mol^{-1}.

The chloroalkanes [6] obey the same general rules as their parent hydrocarbons, eq. (A.9), that is

$$ZPE + (H_T - H_0) = 25.27 + 18.213(n_C - 1) - 4.98(n_{Cl} - 1) \qquad (A.15)$$

where n_{Cl} is the number of chlorine atoms. Although the possible verifications are presently limited in number, in essence because the vibrational

TABLE. A.3. ZPE + $(H_T - H_0)$ energies
of selected chloroalkanes, kcal mol^{-1}

Molecule	ZPE + $(H_T - H_0)$		
	Estd.	Exptl.	
Chloromethane	25.27	25.58	[19]
Dichloromethane	20.29	20.71	[19]
Chloroform	15.31	15.56	[19]
Carbon tetrachloride	10.33	10.02	[19]
Chloroethane	43.48	43.41	[25]
Chloropropane	61.70	61.38	[26]
Chlorobutane	79.91	79.61	[25]

analyses of haloalkanes are unfortunately often incomplete, it appears that
(A.15) is reasonably accurate (Table A.3).

The ZPE + $(H_T - H_0)$ energies of amines and hydrazines also obey simple
additivity rules [6] (Table A.4). For the amines, one can predict their ZPE +
$(H_T - H_0)$ energies from that of NH_3 (22.94 kcal mol^{-1}) and the appropriate
alkane, eq. (A.9), by subtracting 11.55 kcal mol^{-1} for each pair of H atoms
lost during during the condensation giving the amine. Dimethylamine, for
example, is estimated by taking twice the value of CH_4 plus that of NH_3,
less twice 11.55 kcal mol^{-1}. We proceed in similar fashion with the alkyl-
substituted hydrazines. This approximation is generally satisfactory. As
regards diethylamine, it must be taken into account that no provision was
made in our estimate for the (at least partial) hindrance of internal rotations
in the ethyl groups as a result of steric crowding. As indicated in the next
Subsection, detailed analyses reveal a systematic lowering of ZPE + $(H_T -
H_0)$ by ~ 0.3 kcal mol^{-1} for one hindered internal rotation, suggesting that
our estimate for diethylamine is probably too high by ~ 0.6 kcal mol^{-1}.

Nonaromatic cyclic molecules

Cycloalkanes constructed from six-membered rings behave in essence like
their noncyclic counterparts, eq. (A.9), as regards their vibrational energies,
provided that the suppression of internal rotations is adequately taken into
account [5]. Their ZPE + $(H_T - H_0)$ energies can be tentatively evaluated
by counting 18.213 kcal mol^{-1} for each C atom of the molecule (*rule 1*) and
subtracting 0.343 kcal mol^{-1} for each tertiary C atom that is formed (*rule
2*). A number of H atoms are eliminated in the formation of polycyclic com-

TABLE A.4. ZPE + $(H_T - H_0)$ energies of
selected amines and hydrazines, kcal mol^{-1}

Molecule	ZPE + $(H_T - H_0)$	
	Estd.	Exptl.
Ammonia	(22.94)	22.94 [19]
Methylamine	41.08	41.48 [27]
Ethylamine	59.30	59.41 [28]
Propylamine	77.51	77.62 [28]
Isopropylamine	77.17	77.20 [28]
tert-Butylamine	95.38	95.21 [28]
Dimethylamine	59.22	59.44 [29]
Diethylamine	95.65	94.94 [30]
Trimethylamine	77.37	77.76 [31]
Hydrazine	(33.94)	33.94 [32]
Methylhydrazine	52.08	52.00 [33]
1,1-Dimethylhydrazine	69.88	69.98 [34]
1,2-Dimethylhydrazine	70.22	70.20 [35]

pounds, e.g., 2 H atoms in *trans*-decalin. The 11.479 kcal mol^{-1} term of eq.
(A.9) roughly represents the contribution of two H atoms. Accordingly, we
approximate the decrease in ZPE + $(H_T - H_0)$ due to the loss of one H atom
by subtracting 5.74 kcal mol^{-1} per H atom that is removed (*rule 3*). These
rules clearly express the fact that up to this point cyclic compounds are
considered exactly on the same basis as the acyclic ones, following strictly
the description given in (A.9). Considering now that the latter describes
noncyclic molecules in which internal rotations are as free as possible while,
on the other hand, internal rotations are hindered in cyclic structures, we
subtract tentatively $RT/2 = 0.296$ kcal mol^{-1} (at 298.15 K) for each CC
bond in the cycle (*rule 4*). These simple rules, modelled on eq. (A.9), lead
to satisfactory estimates of ZPE + $(H_T - H_0)$ energies for six-membered
cycloalkanes. Of course, there is no reason to assume that the individual
energy contributions in cycloalkanes are necessarily exactly the same as in
noncyclic paraffins. To begin with, the comparison of results obtained for
isomer pairs differing by the number of butane-*gauche* interactions (e.g., *cis*-
vs. *trans*-decalin) suggests [2] a lowering of ZPE + $(H_T - H_0)$ energy by
0.853 kcal mol^{-1} for one *gauche* interaction. A refined analysis indicates,
first, a contribution of 18.249 kcal mol^{-1} for each carbon atom and, second,
a decrease of 0.322 kcal mol^{-1} for each tertiary C atom of the molecule.

TABLE A.5. ZPE + $(H_T - H_0)$ energies of cycloalkanes

Molecule	ZPE + $(H_T - H_0)$, kcal mol^{-1}		
	Rules	(A.16)	Exptl.[a]
Cyclohexane	107.50	107.72	107.54
Methylcyclohexane	125.37	125.65	125.73
trans-Decalin	166.71	166.74	166.90
Adamantane	154.25	153.95	153.54
Bicyclo[2.2.2]octane	130.87	130.83	130.83

[a]The value for bicyclo[2.2.2]octane is given in [36]. The other results are from [37], using the data of [9] and [8].

Moreover, the removal of one H atom in the formation of polycyclic structures reduces the ZPE + $(H_T - H_0)$ energy by 5.925 kcal mol^{-1}. Finally, this energy is further reduced by 0.296 (= $RT/2$) kcal mol^{-1} for each CC bond of the cycle(s). These results, which resemble closely the rules inferred from noncyclic alkanes, are expressed by eq. (A.16)

$$ZPE + (H_T - H_0) = 11.850(1 - m) + 18.249n_{C,cycle} - 0.322n_{tert}$$
$$-(n_{C,cycle} + m - 1)RT/2 - 0.853n_g \qquad (A.16)$$

where $n_{C,cycle}$ and $(n_{C,cycle} + m - 1)$ are, respectively, the numbers of C atoms and CC bonds in the cycle(s) and $2(m - 1)$ is the number of H atoms removed in the formation of polycyclic molecules. The number of butane-*gauche* interactions is n_g. It is noteworthy that the change from chair to boat cyclohexane has no effect (or only a very minor one) on ZPE + $(H_T - H_0)$ [2, 39]. Selected examples are presented in Table A.5.

Any shrinking of a cycle accompanying the removal of one CH_2 group translates into a regular decrease of ZPE + $(H_T - H_0)$ by 18.545 kcal mol^{-1}. It is remarkable that this regularity includes the shrinking from cyclopropane to give the 'two-membered cycle' ethylene (Table A.6)[2].

Similar regularities are also observed [6] for cycles containing one or more oxygen atoms (Table A.7). It suffices to calculate the ZPE + $(H_T - H_0)$ value of the corresponding cycloalkane and to subtract 14.50 kcal mol^{-1} for each CH_2 that is replaced by an oxygen atom. The agreement with spectroscopic results is remarkable considering that this verification includes

---

[2]The fundamental frequencies of cycloheptane are given in [38]. The other results are reported in [2].

TABLE A.6. ZPE + $(H_T - H_0)$ values of
selected cycloalkanes, kcal mol^{-1}

Molecule	ZPE + $(H_T - H_0)$	
	Predicted	Exptl.
Cycloheptane	126.09	124.15
Cyclohexane		107.54
Cyclopentane	89.00	89.0
Cyclobutane	70.45	70.4
Cyclopropane	51.91	51.9
Ethylene	33.36	33.36

an eight-membered cycle with four oxygen atoms as well as small cycles such as ethylene oxide and the 'two-membered cycle' formaldehyde.

Finally, the replacement of CH_2 by NH also manifests itself by a significant lowering of its ZPE + $(H_T - H_0)$ energy [6]. The few spectroscopic analyses which are presently available suggest an order of magnitude of ~ 7.1 kcal mol^{-1} for this lowering. The values thus estimated for piperidine, pyrolidine and ethylene imine are 100.4, 81.9 and 44.8 kcal mol^{-1}, respectively, but are only in moderate agreement with their experimental counterparts, namely, 100.53 for the axial and 100.25 kcal mol^{-1} for the equatorial piperidine [45], ~ 81 kcal mol^{-1} estimated for pyrolidine (from an incomplete spectral analysis [46]), and 45.95 kcal mol^{-1} for ethylene imine [47].

TABLE A.7. ZPE + $(H_T - H_0)$ energies
of oxygen containing cycles, kcal mol^{-1}

Molecule	ZPE + $(H_T - H_0)$		
	Estd.	Exptl.	
Tetrahydropyran	93.04	92.87	[40]
Tetrahydrofuran	74.50	74.61	[40]
Oxetane	55.95	56.14	[41]
Ethylene oxide	37.41	37.25	[42]
Formaldehyde	18.86	18.62	[43]
1,4-Dioxane	78.54	78.48	[40]
1,3,5-Trioxane	64.04	64.4	[44]
1,3,5,7-Tetroxocane	86.63	86.2	[44]

Aromatic molecules

Adequate vibrational energy data are particularly scarse for the series of benzenoid hydrocarbons. Fortunately, it is now justifiable to take advantage from the extraordinary regularity observed for the build-up of alkyl chains, namely, the gain of 18.213 kcal mol^{-1} for each added CH_2 group. Hence it appears safe to use the following formula for alkyl substitution, based on the experimental ZPE + $(H_T - H_0)$ value (66.22 kcal mol^{-1}) calculated for benzene in the harmonic oscillator approximation [2, 19]

$$\text{ZPE} + (H_T - H_0) = 66.22 + 18.21n - 0.343 n_{br} \qquad (A.17)$$

where n is the number of alkyl carbon atoms. In similar fashion, one can treat molecules like 1,2,3,4-tetrahydronaphtalene **1**, or 9,10-dihydroanthracene **2**,

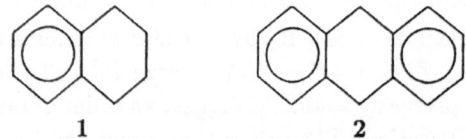

1   **2**

but in these situations attention must be given to the lowering of ZPE + $(H_T - H_0)$ by 11.5 kcal mol^{-1} accompanying the loss of a pair of hydrogen atoms and by $RT/2 = 0.296$ kcal mol^{-1} for each hindered internal rotation in cyclic structures. 1,2,3,4-Tetrahydronaphtalene, for example, gives 66.22 + $4 \times 18.21 - 11.5 - 5 \times 0.296 = 126.1$ kcal mol^{-1}. The addition of two fragments involves a correction of $4RT = 2.37$ kcal mol^{-1} because this term is included in the $(H_T - H_0)$ part of each molecule used as fragment and should not be counted twice in the final product. For example, the ZPE + $(H_T - H_0)$ value of 9,10-dihydroanthracene is estimated from two benzene molecules plus two CH_2 groups ($2 \times 66.22 + 2 \times 18.21$) less 2×11.5, less $4RT$, and less $4 \times RT/2$. This type of estimate is considered to carry an uncertainty not exceeding 0.2 kcal mol^{-1} [48] but evidently additional information is required to asses this matter. Additivity rules described in [48] can be applied with confidence to construct the ZPE + $(H_T - H_0)$ energies of polycyclic aromatic hydrocarbons. For example, using the result ZPE + $(H_T - H_0) = 94.90$ kcal mol^{-1} for naphtalene, deduced from its vibrational spectrum [49], we add to it the difference, 28.68 kcal mol^{-1}, between naphtalene and benzene, thus obtaining an estimate of 123.6 kcal mol^{-1} for anthracene. The same procedure is used for the higher homologues. Finally, using the fundamental frequencies of pyrene [50], it is found that ZPE + $(H_T - H_0) = 133.05$ kcal

TABLE A.8. ZPE + $(H_T - H_0)$ energies
of selected azines, kcal mol^{-1}

Molecule	ZPE + $(H_T - H_0)$		
	Estd.	Exptl.	
Benzene	66.06	66.22	[19]
Pyridine	57.61	57.29	[53]
1,3-Diazine	49.16	49.40	[54]
1,4-Diazine	49.16	49.07	[54]
1,2-Diazine	49.16	49.16	[54]
1,3,5-Triazine	40.71	42.9	[55]

mol^{-1}, and for styrene one obtains 85.80 kcal mol^{-1} from its vibrational spectrum [51]. The zero-point energy of graphite was calculated [52] using Debye's theory, ZPE = 3.68 kcal mol^{-1}.

The loss of vibrational energy accompanying the replacement of an aromatic CH group by N was also investigated [6]: it can be represented by ZPE + $(H_T - H_0) \simeq 66.06 - 8.45 n_N$ kcal mol^{-1}, where n_N is the number of N atoms (Table A.8).

Free radicals

Little is presently known for the free radicals, except for a few alkyl radicals. Their ZPE + $(H_T - H_0)$ energies, deduced from experimental and calculated fundamental frequencies [56], are 20.74 ($CH_3 \cdot$), 39.15 ($C_2H_5 \cdot$) and 74.97 kcal mol^{-1} (tert-$C_4H_9 \cdot$). These data suggest [57] that the ZPE + $(H_T - H_0)$ energies of alkyl radicals R· are systematically lower by 8.85 kcal mol^{-1} than those of their parent hydrocarbons RH.

A.3 Concluding remarks

Simple additivity rules relating ZPE + $(H_T - H_0)$ energies to structural features have proven their usefulness in the past. Here they were primarily examined because of the need to link thermochemical data to theoretical calculations made for molecules in their hypothetical vibrationless state at 0 K. While, of course, preference is given to verifications involving only genuine experimental data, well established structure dependent regularities of zero-point plus heat content energies considerably augments the number of molecules that can be tested. Beyond these stringent needs, it now appears that the arsenal of simple additivity rules can be extended, giving access to additional classes of compounds amenable to simple evaluations. While certainly pleasing in problems requiring their application, the quality of the correlations obtained from *bona fide* spectral analyses at times surpasses what one would normally expect. This is, namely, the case with the remarkable regularities observed for the replacement of a single by a double or a triple carbon–carbon bond, the fundamental similarity between open chain and cyclic alkanes, the progressive shrinking of cycloalkanes and the replacement of CH_2 by O in a cycle. Thus, beyond any consideration of their practical merits, it is certainly interesting to witness the surprising simplicity and the unifying trends governing $ZPE + (H_T - H_0)$ energies in relation to structural features although, admittedly, all of it is presently recognized only on an empirical basis.

Bibliography

[1] D. R. Stull and G. C. Sinke, *Adv. Chem. Ser.*, **18** (1956).

[2] S. Fliszár, "Charge Distributions and Chemical Effects", Springer-Verlag, New York, 1983.

[3] K. S. Pitzer and E. Catalano, *J. Am. Chem. Soc.*, **78**, 4844 (1956); T. L. Cottrell, *J. Chem. Soc.*, 1448 (1948); K. S. Pitzer, *Chem. Rev.*, **27**, 39 (1940); K. S. Pitzer, *J. Chem. Phys.*, **8**, 711 (1940); K. S. Pitzer, *J. Chem. Phys.*, **5**, 473 (1937).

[4] T. Ot, A. Popowicz, and T. Ishida, *J. Phys. Chem.*, **90**, 3080 (1986).

[5] S. Fliszár and J.-L. Cantara, *Can. J. Chem.*, **59**, 1381 (1981).

[6] S. Fliszár, F. Poliquin, I. Bădilescu, and E. Vauthier, *Can. J. Chem.*, **66**, 300 (1988).

[7] S. Fliszár and G. Cardinal, *Can. J. Chem.*, **62**, 2748 (1984).

[8] F. D. Rossini, "Selected Values of Physical and Thermodynamic Properties of Hydrocarbons and Related Compounds", Carnegie Press, Pittsburgh, PA, 1952.

[9] R. G. Snyder and J. H. Schachtschneider, *Spectrochim. Acta*, **21**, 169 (1965).

[10] B. L. Crawford, J. E. Lancaster, and R. G. Inskeep, *J. Chem. Phys.*, **21**, 678 (1953).

[11] A. J. Baines and J. D. R. Howells, *J. Chem. Soc., Faraday Trans.II*, **69**, 532 (1973).

[12] N. Sheppard, *J. Chem. Phys.*, **17**, 74 (1949).

[13] Yu. N. Panchenko, *Spectrochim. Acta*, **31A**, 1201 (1975).

[14] D. A. C. Compton, W. O. George, and W. F. Maddams, *J. Chem. Soc., Perkin II*, 1311 (1977).

[15] D. A. C. Compton, W. O. George, and W. F. Maddams, *J. Chem. Soc., Perkin II*, 1666 (1976); V. A. Kuznetsov, S. Dzhessati, A. R. Kyazimova, and V. I. Tyulin, *Moscow Univ. Chem. Bull.*, **26**(1), 29 (1971), reported by C. W. Bock, Yu. N. Panchenko, S. V. Krasnoshchiokov, and R. Aroca, *J. Mol. Struct.*, **160**, 337 (1987).

[16] Yu. N. Panchenko, V. I. Mochalov, P. Czászár, F. Török, E. Benedetti, and M. Aglietto, *Acta Chim. Hung.*, **114**, 149 (1983); J. R. Durig and D. A. C. Compton, *J. Phys. Chem.*, **83**, 2879 (1979).

[17] C. W. Bock, Yu. Panchenko, S. V. Krasnoshchiokov, and V. Pupyshev, *J. Mol. Struct.(Theochem)*, **148**, 131 (1986).

[18] B. S. Hudson, B. E. Kohler, and K. Schulter, "Excited States", vol. 6, E. C. Lim (ed.), Academic Press, New York, 1982, pp 1–95; R. J. Hemley, B. R. Brooks, and M. Karplus, *J. Chem. Phys.*, **85**, 6550 (1986).

[19] G. Herzberg, "Molecular Spectra and Molecular Structure. II. Infrared and Raman Spectra of Polyatomic Molecules", Van Nostrand Reinhold Co., New York, 1968.

[20] G. A. Crowder and H. Fick, *J. Mol. Struct.*, **147**, 17 (1986).

[21] P. Cossee and J. H. Schachtschneider, *J. Chem. Phys.*, **44**, 97 (1966).

[22] Z. Buric and P. J. Krueger, *Spectrochim. Acta*, **30A**, 2069 (1974).

[23] K. Hamada and H. Morishita, *Z. Physik. Chem.*, **97**, 295 (1975).

[24] R. G. Snyder and G. Zerbi, *Spectrochim. Acta*, **23A**, 391 (1967).

[25] R. G. Snyder and J. H. Schachtschneider, *J. Mol. Spectrosc.*, **30**, 290 (1969).

[26] K. Tanabe and S. Saëki, *J. Mol. Struct.*, **27**, 79 (1975).

[27] Y. Hamada, N. Tanaka, Y. Sugawara, A. Hirakawa, M. Tsuboi, S. Kato, and K. Morokuma, *J. Mol. Spectrosc.*, **96**, 313 (1982).

[28] D. W. Scott, *J. Chem. Thermodyn.*, **3**, 843 (1971).

[29] G. Gamer and H. Wolff, *Spectrochim. Acta*, **29A**, 129 (1973).

[30] A. L. Verma, *Spectrochim. Acta*, **27A**, 2433 (1971).

[31] J. N. Gayles, *Spectrochim. Acta*, **23A**, 1521 (1967).

[32] P. A. Giguère and J. D. Liu, *J. Chem. Phys.*, **20**, 136 (1952).

[33] J. R. Durig and W. C. Harris, *J. Chem. Phys.*, **50**, 1457 (1969).

[34] J. R. Durig and W. C. Harris, *J. Chem. Phys.*, **51**, 4463 (1969).

[35] J. R. Durig and W. C. Harris, *J. Chem. Phys.*, **55**, 1745 (1971).

[36] R. H. Boyd, S. N. Sanwal, S. Chary–Tehany, and D. M. McNally, *J. Phys. Chem.*, **75**, 1264 (1971).

[37] H. Henry, G. Kean, and S. Fliszár, *J. Am. Chem. Soc.*, **99**, 5889 (1977).

[38] D. F. Bocian and H. L. Strauss, *J. Am. Chem. Soc.*, **99**, 2866 (1977).

[39] J. P. Huvenne, G. Vergoten, G. Fleury, S. Odiot, and S. Fliszár, *Can. J. Chem.*, **60**, 1347 (1982).

[40] A. T. Balaban, S. Bădilescu, and I. Bădilescu, "Infrared Spectra of Heterocyclic Compounds" *in* "Physical Methods in Heterocyclic Chemistry", R. R. Gupta (ed.), John Wiley & Sons, New York, 1984.

[41] Gy. Bánhegyi, P. Pulay, and G. Fogarasi, *Spectrochim. Acta*, **39A**, 761 (1983).

[42] N. W. Cant and W. Armstead, *Spectrochim. Acta*, **31A**, 839 (1975).

[43] D. E. Reisner, R. W. Field, J. L. Kinsay, and H. L. Dai, *J. Chem. Phys.*, **80**, 5968 (1984).

[44] M. Kobayashi, *J. Chem. Phys.*, **66**, 32 (1977).

[45] Kh. Kh. Muldagaliev and J. S. Ignatev, *Z. Fiz. Khim.*, **57**, 1508 (1983).

[46] J. C. Evans and J. C. Wahr, *J. Chem. Phys.*, **31**, 661 (1959).

[47] H. T. Hoffman, Jr., G. E. Evans, and J. Glockler, *J. Am. Chem. Soc.*, **73**, 3030 (1951).

[48] S. Fliszár, G. Cardinal, and N. A. Baykara, *Can. J. Chem.*, **64**, 404 (1986).

[49] S. S. Mitra and H. J. Bernstein, *Can. J. Chem.*, **37**, 553 (1959).

[50] A. Bree, R. A. Kydd, T. N. Misra, and V. V. B. Vilkos, *Spectrochim. Acta*, **A27**, 2315 (1971).

[51] A. Marchand and J.-P. Quintard, *Spectrochim. Acta*, **A36**, 941 (1980).

[52] A. Peluso and S. Fliszár, *Can. J. Chem.*, **66**, 2631 (1988).

[53] D. P. Dilalla and H. Stidham, *J. Raman Spectrosc.*, **9**, 103 (1980).

[54] R. C. Lord, A. L. Marston, and F. A. Miller, *Spectrochim. Acta*, **9**, 118 (1957).

[55] W. Pyckhout, I. Collaerts, C. Van Alsenoy, H. J. Geise, A. Almenningen, and R. Seip, *J. Mol. Struct.*, **147**, 321 (1986).

[56] A. Snelson, *J. Chem. Phys.*, **74**, 537 (1970); J. Pacansky and B. Schrader, *J. Chem. Phys.*, **78**, 1033 (1983); B. Schrader, J. Pacansky, and U. Pfeiffer, *J. Phys. Chem.*, **88**, 4069 (1984).

[57] S. Fliszár and C. Minichino, *Can. J. Chem.*, **65**, 2495 (1987).

Index

Editorial Policy

This series aims to report new developments in chemical research and teaching - quickly, informally and at a high level. The type of material considered for publication includes:

1. Preliminary drafts of original papers and monographs
2. Lectures on a new field, or presenting a new angle on a classical field
3. Seminar work-outs
4. Reports of meetings, provided they are
   a) of exceptional interest and
   b) devoted to a single topic.

Texts which are out of print but still in demand may also be considered if they fall within these categories.

The timeliness of a manuscript is more important than its form, which may be unfinished or tentative. Thus, in some instances, proofs may be merely outlined and results presented which have been or will later be published elsewhere. If possible, a subject index should be included. Publication of Lecture Notes is intended as a service to the international chemical community, in that a commercial publisher, Springer-Verlag, can offer a wider distribution to documents which would otherwise have a restricted readership. Once published and copyrighted, they can be documented in the scientific literature.

Manuscripts

Manuscripts should comprise not less than 100 and preferably not more than 500 pages. They are reproduced by a photographic process and therefore must be typed with extreme care. Symbols not on the typewriter should be inserted by hand in indelible black ink. Corrections to the typescript should be made by pasting the amended text over the old one, or by obliterating errors with white correcting fluid. Authors receive 50 free copies and are free to use the material in other publications. The typescript is reduced slightly in size during reproduction; best results will not be obtained unless the text on any one page is kept within the overall limit of 18 x 26.5 cm (7 x $10^{1}/_{2}$ inches). The publishers will be pleased to supply on request special stationary with the typing area outlined.

Manuscripts should be sent to one of the editors or directly to Springer-Verlag, Heidelberg.

Lecture Notes in Chemistry

For information about Vols. 1–22
please contact your bookseller or Springer-Verlag